CORDYCEPS MILITARIS CULTIVATION: A PRACTICAL GUIDE.

First edition. August 26, 2026.

ISBN: 979-8235419896

Written by Pragyan Ray and Pradip Kumar Ray.

Table of Contents

Cordyceps militaris Cultivation: A Practical Guide.

Pragyan Ray

&

Pradip Kumar Ray

Copyright

Cordyceps militaris Cultivation: A Practical Guide

Second Edition

Published by: **PKRBUR PUBLICATION** , A Platform for Emerging Writers , Burdwan, West Bengal, India

Website: https://pkrburpub.com[1]

Author: Pragyan Ray & **Pradip Kumar Ray**

Manuscript Reader: **Babli Roy**

Editor: **Sonali Roy**

---

1. https://pkrburpub.com/

ISBN: 978-81-686880-0-1
Printed in India
Page: 242

ISBN 978-81-968860-0-[illegible]

Printed in India

Page: 242

The publisher and the author have made every effort to ensure the accuracy of the information contained in this book at the time of publication. However, neither the author nor the publisher assumes responsibility for any errors, omissions, or consequences arising from the use of the information contained herein.

Preface

It gives us immense pleasure to present the another book on cordyceps titled as "*Cordyceps militaris Cultivation: A Practical Guide*". Since the publication of the first book on cordycepes , scientific research, cultivation technologies, controlled-environment agriculture, and commercial applications of *Cordyceps militaris* have advanced considerably. At the same time, interest among researchers, students, entrepreneurs, farmers, and biotechnology professionals has grown rapidly across the world.

The overwhelming response received from readers of the first book on cordyceps encouraged us to prepare this thoroughly revised and expanded edition. Rather than merely updating existing chapters, this edition has been substantially rewritten with new scientific information, improved cultivation techniques, enhanced laboratory protocols, practical troubleshooting guides, quality management systems, commercialization strategies, and recent developments in biotechnology.

This book has been designed as a practical handbook rather than only a theoretical text. Every chapter has been prepared with the objective of helping readers understand not only *how* to cultivate *Cordyceps militaris*, but also *why* each step is scientifically important. Special emphasis has been placed on laboratory hygiene, aseptic techniques, substrate preparation, environmental management, contamination control, post-harvest handling, value addition, entrepreneurship, and quality assurance.

The also introduces several new features, including detailed Standard Operating Procedures (SOPs), laboratory record formats, troubleshooting charts, business planning, quality control systems, commercialization strategies, and future prospects of the industry. These additions make the book useful not only for beginners but also for researchers, commercial growers, universities, biotechnology laboratories, training institutes, and entrepreneurs.

While every effort has been made to present scientifically accurate and practical information, mushroom cultivation is influenced by numerous variables such as strain characteristics, environmental conditions, equipment quality, and cultivation practices. Therefore, readers are encouraged to conduct small-scale trials before undertaking commercial production and to continue learning through research and practical experience.

We sincerely hope that this book will contribute to the advancement of medicinal mushroom cultivation and inspire more individuals to participate in this exciting field of biotechnology.

We welcome constructive suggestions and feedback from readers for further improvement of future editions.

Pragyan Ray & Pradip Kumar Ray

PRAGYAN RAY AND PRADIP KUMAR RAY

Author

Manuscript Reader's Impression

The of *Cordyceps militaris Cultivation: A Practical Guide* is a remarkable improvement over the previous edition. The manuscript successfully combines scientific accuracy with practical experience, making it suitable for readers ranging from beginners to professional cultivators and researchers.

One of the major strengths of this edition is its systematic presentation. Each chapter progresses logically from basic biological concepts to advanced cultivation technologies, laboratory management, quality assurance, commercialization, and future research opportunities. The inclusion of troubleshooting guides, Standard Operating Procedures (SOPs), laboratory record formats, business planning, and quality management systems significantly enhances its practical value.

The author has taken considerable care to present scientific information in a clear and accessible manner without sacrificing technical accuracy. Complex concepts have been explained in simple language, supported by practical examples and step-by-step guidance. The emphasis on laboratory discipline, documentation, food safety, and ethical commercialization reflects a mature understanding of modern medicinal mushroom production.

This book will serve as a valuable reference for students, researchers, mushroom growers, entrepreneurs, biotechnology professionals, universities, and training institutions. It is likely to remain a useful resource for many years because it bridges the gap between academic knowledge and practical application.

I congratulate the author on producing a comprehensive and informative work and wish this every success.

— Babli Roy

Manuscript Reader

Acknowledgement

The completion of this has been possible through the encouragement, support, and inspiration of many individuals and institutions.

I express my deepest gratitude to Almighty God for granting me the strength, patience, and determination to complete this work. My sincere thanks go to my wife, Mrs. Sonali Roy, for her constant encouragement, understanding, and unwavering support throughout the preparation of this book. Her confidence in my work has always been a source of motivation.

I am grateful to Ms. Babli Roy, Manuscript Reader, for her careful reading of the manuscript and for offering valuable suggestions that greatly improved the clarity and quality of this edition.

I also acknowledge the contributions of researchers, scientists, universities, and biotechnology laboratories around the world whose published research has enriched our understanding of *Cordyceps militaris*. Their work has provided a strong scientific foundation for this book.

Special thanks are extended to readers of the first book whose valuable comments and suggestions inspired us many of the improvements incorporated into this revised book with enlargement.

Finally, we express my heartfelt gratitude to our family, friends, well-wishers, and everyone who encouraged us during the preparation of this book.

—Pragyan Ray & Pradip Kumar Ray

Dedication

This book is lovingly dedicated to my beloved parents, Late Shri Monoranjan Ray and Late Smt. Anima Ray, whose love, values, sacrifices, and blessings continue to inspire every step of my life. I also dedicate this book to all students, researchers, mushroom cultivators, entrepreneurs, and lifelong learners who believe that scientific knowledge, perseverance, and innovation can create a better future.

— Pradip Kumar Ray

Surrender

Whatever knowledge we have acquired, whatever experience we have gained, and whatever work has been accomplished through this book are ultimately offerings at the feet of the Almighty. With humility, gratitude, and devotion, we surrender this work to God, praying that it may become a source of knowledge, inspiration, and benefit to humanity. May this humble effort contribute to scientific progress, ethical entrepreneurship, and the well-being of all.

"May knowledge illuminates every mind, and may wisdom guide every action."

— Pragyan Ray & Pradip Kumar Ray

Disclaimer

This book is intended solely for educational, scientific, and informational purposes. The cultivation methods, laboratory procedures, business guidance, and technical information presented herein are based on scientific literature, practical experience, and generally accepted cultivation practices available at the time of publication.

The author and publisher do not guarantee specific cultivation results, production yields, profitability, or commercial success. Outcomes may vary depending on fungal strain, substrate formulation, environmental conditions, equipment, operator skill, and numerous other factors beyond the control of the author and publisher.

Any discussion of biological activities, traditional uses, or scientific research relating to *Cordyceps militaris* should not be interpreted as medical advice or as evidence that the mushroom can diagnose, treat, cure, or prevent any disease. Readers should consult qualified healthcare professionals regarding any medical concerns.

Readers are solely responsible for complying with all applicable laws, regulations, licensing requirements, food safety standards, environmental regulations, and commercial practices in their respective countries.

Neither the author nor the publisher shall be liable for any direct, indirect, incidental, or consequential loss, damage, or injury arising from the use or misuse of the information contained in this publication.

Important Precautions

To ensure safe and successful cultivation of *Cordyceps militaris*, readers are advised to observe the following precautions:

1. Always maintain strict laboratory hygiene and aseptic techniques.
2. Use authenticated cultures obtained from reliable sources.
3. Sterilize all media, equipment, and cultivation containers properly before use.
4. Wear appropriate personal protective equipment (PPE), including gloves, masks, and laboratory coats, during culture work.
5. Never consume contaminated cultures or unidentified mushrooms.
6. Dispose of contaminated materials safely in accordance with local biosafety and environmental regulations.
7. Keep accurate records of all cultivation activities, environmental conditions, and production batches.

8. Calibrate laboratory equipment periodically to ensure accurate measurements.
9. Conduct pilot-scale cultivation before investing in commercial production.
10. Follow all applicable food safety, occupational safety, and regulatory requirements for cultivation, processing, packaging, labeling, and marketing.
11. Avoid making unsupported health or therapeutic claims regarding *Cordyceps militaris* products.
12. Continuously update your knowledge through scientific literature, training programs, and practical experience.

Careful adherence to these precautions will improve production consistency, reduce contamination, enhance product quality, and promote safe, responsible, and sustainable cultivation practices.

—Publisher & Author

PART-I

THEORETICAL CULTIVATION OF CORDYCEPS MILITARIS

Chapter 1: Introduction to Medicinal Mushrooms

Medicinal mushrooms have fascinated humankind for thousands of years. Long before the development of modern pharmaceuticals, civilizations across Asia, Europe, and other parts of the world relied on mushrooms not merely as food but also as valuable natural remedies. Ancient healers observed that certain mushrooms appeared to improve vitality, increase endurance, accelerate recovery from illness, and support long-term health. These observations, passed down through generations, laid the foundation for the scientific interest in medicinal mushrooms that continues today.

Unlike ordinary edible mushrooms that are consumed primarily for their nutritional value, medicinal mushrooms contain a wide range of biologically active compounds capable of influencing various physiological processes in the human body. Modern research has identified polysaccharides, beta-glucans, nucleosides, triterpenoids, peptides, sterols, antioxidants, vitamins, minerals, and numerous secondary metabolites that contribute to their medicinal properties. These compounds are being investigated for their potential roles in supporting immune function, reducing oxidative stress, regulating inflammation, improving metabolism, and promoting overall wellness.

The growing popularity of natural health products has renewed worldwide interest in medicinal fungi. Consumers are increasingly seeking plant- and fungus-based alternatives that may complement healthy lifestyles. As scientific research advances, medicinal mushrooms are finding applications in nutritional supplements, functional foods, sports nutrition, cosmetics, biotechnology, and pharmaceutical research. Universities, research institutes, and biotechnology companies across the world continue to explore their chemistry, genetics, cultivation methods, and possible therapeutic applications.

Medicinal mushrooms are remarkably diverse. Among the best-known species are *Ganoderma lucidum* (Reishi), traditionally associated with longevity and immune support; *Lentinula edodes* (Shiitake), valued both as food and for its bioactive polysaccharides; *Hericium erinaceus* (Lion's Mane), which has attracted attention for its potential effects on cognitive health; *Trametes versicolor* (Turkey Tail), extensively studied for immune-related compounds; and *Grifola frondosa* (Maitake), recognized for its nutritional and medicinal value. Each species possesses its own unique composition and characteristics.

Among all these remarkable fungi, *Cordyceps* occupies a particularly special position. Unlike most mushrooms that grow on decaying wood or soil, many species of *Cordyceps* have evolved extraordinary ecological relationships with insects. Their unusual life cycle has fascinated scientists for centuries and inspired countless legends in traditional medicine. Today,

Cordyceps militaris has become one of the most extensively cultivated medicinal fungi because it can be produced under controlled conditions while containing valuable bioactive compounds such as cordycepin and adenosine.

The increasing commercial cultivation of *Cordyceps militaris* has transformed what was once an extremely rare natural resource into a practical crop suitable for research laboratories, educational institutions, entrepreneurs, and even dedicated hobbyists. Advances in microbiology, tissue culture, sterilization technology, environmental control, and cultivation techniques now make it possible to grow high-quality fruiting bodies throughout the year with appropriate knowledge and equipment.

Nevertheless, successful cultivation requires far more than simply following a recipe. It demands an understanding of fungal biology, aseptic techniques, environmental management, quality control, contamination prevention, and post-harvest handling. Many beginners become discouraged after encountering contamination, poor fruiting, or disappointing yields because they lack an understanding of the underlying biological principles. This book aims to bridge that gap by combining scientific explanations with practical, step-by-step guidance.

Throughout this second edition, readers will learn not only how *Cordyceps militaris* grows, but also why each cultivation step is important, how environmental conditions influence development, how to recognize problems before they become serious, and how to produce consistently high-quality mushrooms. Scientific evidence is presented in a balanced manner, distinguishing well-established findings from promising areas that still require further research.

Whether you are a student, researcher, mushroom enthusiast, entrepreneur, healthcare professional, or simply curious about one of nature's most fascinating organisms, this book is designed to provide a reliable foundation. It combines biological science, practical cultivation methods, quality assurance, commercial perspectives, and responsible use into a single comprehensive reference.

Chapter Summary

By the end of this chapter, you should understand:

- What medicinal mushrooms are and how they differ from ordinary edible mushrooms.
- Why medicinal mushrooms have become increasingly important in modern health research.
- The major groups of medicinal mushrooms and their general applications.
- Why *Cordyceps militaris* is one of the most important medicinal fungi today.
- How this book is organized to guide you from basic concepts to advanced cultivation and commercial production.

The next chapter will explore the fascinating history, discovery, taxonomy, natural habitat, and global significance of *Cordyceps*, providing the scientific foundation for the practical cultivation techniques described later in the book.

Chapter 2 : Understanding Cordyceps

Nature's Extraordinary Medicinal Fungus

"Among the countless organisms that inhabit our planet, few are as mysterious, biologically fascinating, and medically promising as Cordyceps."

2.1 Introduction

The natural world contains millions of living organisms, yet only a handful have attracted the combined attention of traditional healers, modern scientists, pharmaceutical researchers, and commercial cultivators. **Cordyceps** is one of those extraordinary organisms. Its remarkable life cycle, unique chemistry, and centuries-long reputation as a valuable medicinal fungus have made it one of the most studied fungi in the world.

For centuries, people living in the Himalayan regions regarded Cordyceps as a priceless natural treasure. Shepherds noticed that grazing animals became noticeably more energetic after consuming grasses containing this unusual fungus. Over time, local communities began collecting it for medicinal purposes, believing it enhanced strength, vitality, respiratory health, and longevity.

Today, modern science has revealed that many of these traditional observations are linked to naturally occurring bioactive compounds such as **cordycepin**, **adenosine**, **polysaccharides**, and several other biologically active molecules. Although research continues, Cordyceps has become an important subject in biotechnology, microbiology, pharmacology, nutrition, and functional food research.

This chapter introduces readers to the remarkable world of Cordyceps, explaining what it is, how it differs from ordinary mushrooms, where it grows naturally, and why it has become one of the most commercially important medicinal fungi.

2.2 What is Cordyceps?

Cordyceps is a group of **parasitic fungi** belonging to the phylum **Ascomycota**. Unlike common edible mushrooms that grow on wood, soil, or organic matter, many Cordyceps species complete part of their life cycle by infecting insects and other arthropods.

After infecting its host, the fungal mycelium gradually colonizes the insect's body. Under natural conditions, the fungus eventually produces a fruiting body that emerges from the host and releases spores to begin the next generation. Although this life cycle may appear unusual, it represents a highly specialized ecological adaptation that has evolved over millions of years.

Fortunately, the species **Cordyceps militaris**, which is the primary focus of this book, can now be cultivated without relying on wild insect hosts. Modern cultivation techniques allow researchers and growers to produce fruiting bodies on carefully formulated nutrient substrates under controlled laboratory conditions. This advancement has made Cordyceps cultivation safer, more sustainable, and commercially viable.

2.3 Meaning of the Name "Cordyceps"

The name **Cordyceps** originates from two Latin words:

- **Cord** meaning "club"
- **Ceps** meaning "head"

The mature fruiting body resembles a small orange or yellow club emerging from the substrate or insect host.

This distinctive appearance inspired the scientific name that continues to be used today.

2.4 Why is Cordyceps So Special?

Cordyceps differs from ordinary mushrooms in several important ways.

Unlike button mushrooms, oyster mushrooms, or shiitake mushrooms, Cordyceps produces several unique compounds that are relatively rare in nature.

These include:

- Cordycepin
- Adenosine
- D-mannitol
- Polysaccharides
- Ergosterol
- Various peptides
- Essential amino acids
- Phenolic compounds
- Antioxidants

Many of these compounds are currently being investigated for their biological activities in laboratory and clinical research.

The fungus is also remarkable because of its:

- unusual life cycle
- medicinal history
- commercial value
- pharmaceutical potential
- attractive appearance
- increasing demand worldwide

2.5 Natural Habitat

Wild Cordyceps grows primarily in alpine grasslands located at elevations between **3,000 and 5,500 metres** above sea level.

Major natural habitats include:

- Tibet
- Nepal
- Bhutan
- Qinghai Province
- Sichuan Province
- Yunnan Province
- Parts of northern India

These regions experience:

- low temperatures
- thin atmosphere
- high ultraviolet radiation
- prolonged snow cover
- short growing seasons

Such harsh environments contribute to the rarity and high commercial value of naturally occurring Cordyceps.

2.6 Cordyceps in India

India possesses suitable ecological regions for wild Cordyceps, particularly within the Himalayan belt.

Wild collections have been reported from:

- Uttarakhand
- Himachal Pradesh
- Arunachal Pradesh
- Sikkim

Because of increasing commercial demand and overharvesting, many natural populations have declined significantly. Conservation efforts and sustainable harvesting practices are therefore essential to protect these fragile ecosystems.

Today, most research and commercial production in India focuses on **cultivated**

Cordyceps militaris, which offers a sustainable alternative to harvesting wild specimens.

2.7 Why Has Cordyceps Become So Expensive?

Wild Cordyceps is often referred to as **"Himalayan Gold."** Several factors contribute to its exceptionally high price:

Scarcity

Wild specimens occur only in specific high-altitude environments.

Difficult Collection

Collectors often spend weeks trekking through remote mountain regions.

Short Harvest Season

The fungus can only be collected during a limited period each year.

Growing Global Demand

Demand from pharmaceutical companies, traditional medicine practitioners, and consumers has increased substantially.

Medicinal Reputation

Centuries of traditional use have enhanced its perceived value. These factors have encouraged the development of artificial cultivation methods capable of producing high-quality Cordyceps under laboratory conditions.

2.8 Why Cultivate Cordyceps militaris Instead of Harvesting Wild Cordyceps?

Modern cultivation offers numerous advantages.

Sustainable Production

No pressure is placed on fragile natural ecosystems.

Consistent Quality

Controlled cultivation produces more uniform products.

Lower Cost

Artificial cultivation is far less expensive than collecting wild specimens.

Better Safety

Controlled production reduces contamination risks.

Scientific Research

Researchers can obtain standardized materials for experimentation.

Commercial Opportunity

Cultivation allows entrepreneurs to establish profitable businesses while supporting environmental conservation.

2.9 Current Global Importance

Today, Cordyceps has become an important commodity in several industries. These include:

- Nutraceuticals
- Functional foods
- Dietary supplements
- Traditional medicine
- Biotechnology
- Pharmaceutical research
- Cosmetics

- Sports nutrition
- Wellness products

Several countries have invested heavily in research and commercial production, including China, South Korea, Japan, Thailand, Vietnam, the United States, and increasingly, India. The global market continues to expand as scientific understanding and consumer awareness increase.

2.10 Common Misconceptions

Despite its popularity, many myths surround Cordyceps. Some people believe it is a miracle cure for every disease. This is incorrect. Cordyceps should be viewed as a **medicinal fungus with promising biological activities**, not as a replacement for professional medical care or evidence-based treatments. Research has demonstrated encouraging results in laboratory studies and some clinical investigations, but the strength of evidence varies depending on the condition being studied.

Throughout this book, scientific findings will be presented objectively, distinguishing well-supported evidence from areas where further research is still needed.

Key Takeaways

After completing this chapter, readers should understand:

- What Cordyceps is and how it differs from ordinary mushrooms.
- Why *Cordyceps militaris* is preferred for cultivation.
- The natural distribution and ecological significance of Cordyceps.
- The reasons behind the high market value of wild Cordyceps.
- The importance of sustainable cultivation.
- The major industries that now utilize Cordyceps.
- The need to evaluate health claims using scientific evidence rather than myths.

Chapter 3: History of Cordyceps

From Ancient Himalayan Traditions to Modern Biotechnology

"The history of Cordyceps is not merely the story of a fungus; it is the story of how human curiosity transformed an extraordinary natural phenomenon into one of the world's most valuable medicinal resources."

3.1 Introduction

Throughout history, nature has provided humankind with countless medicinal resources. Many of today's modern medicines originated from plants, fungi, and microorganisms that were first used in traditional healing systems. Among these remarkable natural resources, **Cordyceps** occupies a unique place because its discovery combined careful observation of nature, centuries of traditional knowledge, and modern scientific investigation. Unlike many medicinal mushrooms that were cultivated early in history, Cordyceps remained a mysterious organism for centuries. It was found only in remote alpine regions where harsh climatic conditions limited human access. As a result, knowledge about its properties spread slowly through local communities before eventually reaching physicians, researchers, and scientists around the world. Today, Cordyceps is recognized as one of the most extensively researched medicinal fungi, yet its fascinating journey began high in the Himalayan mountains.

3.2 The Discovery of Cordyceps

Historical records suggest that local shepherds living in the Himalayan plateau were among the first to notice the unusual effects associated with Cordyceps. During the summer months, yak and sheep grazed on alpine grasslands where wild Cordyceps naturally occurred. Shepherds observed that animals consuming these grasses often appeared more energetic, active, and capable of travelling long distances across rugged mountain terrain.

These repeated observations led local communities to investigate the mysterious organism growing among the grasses.

Initially, Cordyceps was collected primarily for personal use. Over generations, it became an important part of traditional healing practices in several Himalayan regions. Although many of these historical accounts are based on oral traditions rather than formal scientific documentation, they illustrate how careful observation of nature often precedes scientific discovery.

3.3 Cordyceps in Traditional Chinese Medicine

Traditional Chinese Medicine (TCM) has a history spanning more than two thousand years. Within this medical system, Cordyceps became highly valued as a natural tonic believed to support vitality and overall well-being.

Historical texts describe its use for conditions involving fatigue, weakness, respiratory discomfort, and recovery after prolonged illness.

It was also regarded as a rare and precious ingredient reserved for special occasions because of its limited availability. In classical TCM, Cordyceps was often combined with other medicinal herbs rather than being used alone. Physicians believed that carefully selected herbal combinations produced balanced therapeutic effects.

While traditional uses provide valuable historical context, it is important to recognize that modern scientific evaluation is necessary to determine which traditional claims are supported by reliable evidence.

3.4 Cordyceps in Tibetan Medicine

Tibetan medicine, known for its close relationship with the Himalayan environment, also incorporated Cordyceps into various traditional formulations. Practitioners regarded it as a valuable natural resource collected from high-altitude grasslands. Because harvesting required considerable effort and exposure to harsh environmental conditions, Cordyceps was treated with great respect. Traditional Tibetan medical literature describes its use as part of broader therapeutic approaches that emphasized diet, lifestyle, herbal medicines, and balance within the body.

3.5 Early Scientific Curiosity

European naturalists first became interested in Cordyceps during the eighteenth and nineteenth centuries when explorers began documenting the unusual biodiversity of Asia. Scientists were fascinated by the fungus because it appeared to emerge directly from insects. This unusual relationship challenged existing understanding of fungal biology. Detailed microscopic observations gradually revealed that Cordyceps was not a plant but a highly specialized fungus capable of infecting insect larvae before producing its characteristic fruiting body. These discoveries marked the beginning of scientific investigations into Cordyceps taxonomy, ecology, and life cycle.

3.6 Taxonomic Development

As fungal classification improved, researchers recognized that the traditional genus *Cordyceps* included a remarkably diverse group of species. Advances in molecular biology and DNA sequencing later demonstrated that several species previously classified within the genus actually belonged to different evolutionary lineages.

Consequently, taxonomists established additional genera, including:

- **Ophiocordyceps**
- **Metacordyceps**
- **Elaphocordyceps**

These revisions improved scientific accuracy while preserving the historical importance of the name "Cordyceps" in both scientific literature and public awareness.

3.7 The Challenge of Limited Supply

For centuries, nearly all Cordyceps available in the marketplace originated from wild collection. This dependence created several problems.

Limited Production

Natural populations produced only small quantities each year.

Difficult Harvesting

Collectors travelled to remote alpine regions under challenging weather conditions.

Environmental Pressure

Increasing demand resulted in excessive harvesting in some locations.

Rising Prices

Scarcity caused market prices to increase dramatically.

These challenges encouraged researchers to explore artificial cultivation methods capable of producing Cordyceps under controlled laboratory conditions.

3.8 The Breakthrough in Artificial Cultivation

One of the most significant milestones in the history of Cordyceps was the successful cultivation of **Cordyceps militaris**. Researchers discovered that this species could produce fruiting bodies on specially prepared nutrient substrates without relying on wild insect hosts.

This breakthrough transformed Cordyceps from an extremely rare natural resource into a commercially cultivable medicinal fungus. Artificial cultivation offered numerous advantages:

- Sustainable production
- Consistent quality
- Reduced environmental impact
- Lower production costs
- Better quality control
- Greater availability for scientific research

Today, most commercially available *Cordyceps militaris* products originate from controlled cultivation rather than wild collection.

3.9 The Rise of Modern Research

Interest in Cordyceps expanded rapidly during the late twentieth and early twenty-first centuries. Scientists from multiple disciplines—including microbiology, pharmacology, biotechnology, food science, and nutrition—began investigating its chemical composition and biological activities. Research focused on topics such as:

- Bioactive compounds
- Cultivation techniques
- Genetic diversity
- Fermentation technology
- Functional foods
- Nutraceutical development
- Quality control
- Environmental sustainability

Although many studies have produced encouraging findings, researchers continue to emphasize the need for well-designed clinical trials to better understand the potential health effects of Cordyceps in humans.

3.10 Cordyceps in the Twenty-First Century

Today, Cordyceps is no longer limited to remote Himalayan communities. It is cultivated in laboratories, universities, research institutes, biotechnology companies, and commercial mushroom farms across many countries. Applications now extend far beyond traditional medicine and include:

- Dietary supplements
- Functional beverages
- Health foods
- Sports nutrition
- Biotechnology research
- Pharmaceutical investigations
- Cosmetics
- Academic education

This remarkable transformation reflects the growing integration of traditional knowledge with modern scientific research.

3.11 Lessons from History

The history of Cordyceps teaches several important lessons. First, careful observation of nature can lead to valuable scientific discoveries. Second, traditional knowledge can provide useful starting points for research, but historical claims should always be evaluated using rigorous scientific methods. Third, sustainable cultivation is essential to protect natural ecosystems while meeting increasing global demand. Finally, continued collaboration between researchers, cultivators, healthcare professionals, and policymakers will play a vital role in ensuring that Cordyceps is developed responsibly for future generations.

Chapter Summary

In this chapter, you learned:

- How Cordyceps was first recognized in the

Himalayan region.

- Its historical role in Traditional Chinese Medicine and Tibetan medicine.
- The scientific discoveries that revealed its unique biology.
- Why artificial cultivation became necessary.
- How biotechnology transformed *Cordyceps militaris* into a commercially important medicinal fungus.
- Why sustainable cultivation and evidence-based research are essential for its future.

Key Terms

- Alpine ecosystem
- Traditional Chinese Medicine (TCM)
- Tibetan Medicine
- Artificial cultivation
- Biotechnology
- Sustainable harvesting
- Fruiting body
- Taxonomy

- Bioactive compounds

Practical Insight

History reminds us that successful cultivation begins with understanding the organism itself. The remarkable journey of Cordyceps—from a rare Himalayan fungus to a globally cultivated medicinal mushroom—demonstrates how science and traditional knowledge can complement each other. As we move into the next chapter, we will examine the biology and life cycle of *Cordyceps militaris*, knowledge that forms the foundation for successful cultivation and quality production.

Chapter 4 : Biology and Life Cycle of Cordyceps militaris

"Successful cultivation begins with understanding biology. A grower who understands how Cordyceps lives, grows, and reproduces can prevent problems before they occur."

Learning Objectives

After completing this chapter, you will be able to:

- Understand the biological structure of *Cordyceps militaris.*
- Identify the different stages of its life cycle.
- Explain how the fungus grows and reproduces.
- Understand the environmental conditions required at each growth stage.
- Apply biological knowledge to improve cultivation success.

4.1 Introduction

Like every living organism, *Cordyceps militaris* follows a well-defined life cycle. Although it appears simple from the outside, its development involves highly organized biological processes. Each stage—from spore germination to fruiting body formation—is influenced by temperature, humidity, nutrients, oxygen, and light.

In nature, *Cordyceps* survives by interacting with insect hosts. However, in modern cultivation, scientists have successfully adapted its growth to artificial nutrient media, allowing fruiting bodies to develop without natural hosts. Understanding these biological principles enables cultivators to replicate favourable conditions and achieve consistent production.

4.2 Classification

The scientific classification of *Cordyceps militaris* is as follows:

Rank	Classification
Kingdom	Fungi
Division	Ascomycota
Class	Sordariomycetes
Order	Hypocreales
Family	Cordycipitaceae
Genus	Cordyceps
Species	*Cordyceps militaris*

Unlike green plants, fungi lack chlorophyll and cannot produce food through photosynthesis. Instead, they absorb nutrients from organic materials in their surroundings.

4.3 Structure of *Cordyceps militaris*

Although mature fruiting bodies appear simple, they consist of several specialized structures.

1. Spores

Spores are microscopic reproductive units that function much like seeds in plants. They are produced in large numbers and dispersed into the environment, where they may germinate under favourable conditions.

2. Hyphae

When a spore germinates, it produces thin, thread-like filaments called **hyphae**.

Each hypha grows by extending at its tip, branching repeatedly as it searches for nutrients.

Hyphae are the basic building blocks of the fungus.

3. Mycelium

A network of interconnected hyphae forms the **mycelium**.

The mycelium is the main vegetative body of the fungus and is responsible for:

- nutrient absorption
- growth
- enzyme production
- substrate colonization

Healthy mycelium is usually white, dense, and evenly distributed throughout the substrate.

4. Fruiting Body

Under suitable environmental conditions, the mycelium develops reproductive structures known as **fruiting bodies**. These upright, club-shaped structures are the part most commonly harvested for medicinal and commercial use. Healthy fruiting bodies are typically bright orange to deep orange, although colour intensity may vary depending on cultivation conditions.

5. Asci and Ascospores

Inside the mature fruiting body are microscopic sac-like structures called **asci**. Each ascus contains several **ascospores**, which are released into the environment to complete the life cycle.

4.4 The Life Cycle of *Cordyceps militaris*

The life cycle can be divided into six major stages.

Stage 1: Spore Release

When the fruiting body reaches maturity, microscopic spores are released into the surrounding air. In natural environments, wind, insects, and environmental movement help disperse these spores over considerable distances. Only a small proportion of released spores eventually land in environments suitable for germination.

Stage 2: Spore Germination

If a spore lands on a suitable nutrient source under favourable environmental conditions, it begins to germinate. Essential requirements include:

- adequate moisture
- suitable temperature
- oxygen
- available nutrients

During germination, the spore develops one or more germ tubes that eventually become hyphae.

Stage 3: Mycelial Growth

The germinating hyphae continue to branch and expand, producing an extensive mycelial network. At this stage:

- nutrients are absorbed
- enzymes are secreted
- substrate is colonized
- energy reserves are accumulated

For cultivators, this is one of the most critical stages. Any contamination by bacteria or competing fungi can significantly reduce success.

Stage 4: Primordia Formation

Once the substrate has been fully colonized and environmental conditions become favourable, the fungus begins forming tiny structures known as **primordia**. These are often referred to as "pins" because of their small, pin-like appearance. Factors encouraging primordia formation include:

- appropriate lighting
- reduced carbon dioxide
- correct humidity
- suitable temperature

Failure to provide these conditions often delays or prevents fruiting.

Stage 5: Fruiting Body Development

The primordia gradually elongate into recognizable fruiting bodies.

During this period:

- rapid cell division occurs
- pigments develop
- bioactive compounds accumulate
- moisture demand increases

Environmental stability is essential. Sudden fluctuations in temperature or humidity may result in malformed fruiting bodies or reduced yields.

Stage 6: Maturation and Spore Production

As the fruiting body matures, asci and ascospores develop. In commercial cultivation, harvesting is generally performed before excessive spore release, as this helps maintain product quality and cleanliness within the growing environment.

4.5 Environmental Requirements During Growth

Different stages of development require different environmental conditions.

Growth Stage	Main Requirement
Spore germination	Moisture and nutrients
Mycelial growth	Stable temperature and cleanliness
Primordia formation	Light, humidity, fresh air
Fruiting	Balanced humidity, light, oxygen
Maturation	Stable environment and monitoring

Successful cultivation depends on maintaining these conditions consistently rather than making frequent adjustments.

4.6 Nutritional Requirements

Like other fungi, *Cordyceps militaris* requires both macronutrients and micronutrients for growth. Important nutrients include:

Carbon Sources

- Brown rice

- Glucose
- Sucrose
- Malt extract

Nitrogen Sources

- Silkworm pupae powder
- Yeast extract
- Peptone
- Soy protein

Minerals

- Potassium
- Magnesium
- Calcium
- Phosphorus
- Iron

Vitamins

Trace amounts of B-complex vitamins contribute to healthy growth and metabolism.

Balanced nutrition supports vigorous mycelial growth and high-quality fruiting body formation.

4.7 Why Contamination Occurs

During cultivation, the nutrient-rich substrate can also support the growth of unwanted microorganisms. Common contaminants include:

- bacteria
- moulds
- wild yeasts

These organisms compete with *Cordyceps* for nutrients and may completely inhibit fruiting. Maintaining strict cleanliness and proper sterilization is therefore one of the most important responsibilities of every cultivator.

4.8 Biological Factors Affecting Yield

Several biological factors determine production success:

- Quality of the mother culture

- Genetic stability
- Age of the culture
- Nutrient composition
- Sterility
- Moisture content
- Oxygen availability
- Light exposure
- Temperature stability

Even a small imbalance in one of these factors may reduce yield or quality.

Practical Tips for Growers

- Always begin with a healthy, uncontaminated culture.
- Avoid exposing sterile cultures to open air unnecessarily.
- Label all culture containers with dates and batch numbers.
- Inspect cultures daily for unusual colours, odours, or growth patterns.
- Record environmental conditions in a cultivation logbook.

- Remember that patience is essential—healthy growth cannot be rushed.

Common Mistakes

◈ Assuming faster growth is always better.
◈ Opening culture containers repeatedly.
◈ Ignoring early signs of contamination.
◈ Using old or weakened cultures.
◈ Allowing large temperature fluctuations.
◈ Overwatering or drying the substrate.

Chapter Summary

In this chapter, you have learned:

- The biological classification of *Cordyceps militaris*.
- The structure and function of spores, hyphae, mycelium, and fruiting bodies.
- The six major stages of its life cycle.
- The environmental conditions required for successful development.
- The nutritional needs of the fungus.

- The importance of contamination prevention.
- The biological factors influencing yield and quality.

Understanding these biological principles provides the scientific foundation for successful cultivation. In the next chapter, we will compare **wild *Cordyceps sinensis*** and **cultivated *Cordyceps militaris***, examining their similarities, differences, cultivation methods, market value, sustainability, and practical importance for growers and consumers. This comparison will help readers appreciate why *C. militaris* has become the preferred species for commercial cultivation and scientific research.

Chapter 5 : Cordyceps sinensis vs. Cordyceps militaris

A Comprehensive Comparison

"Understanding the differences between wild Cordyceps and cultivated Cordyceps is essential for researchers, growers, consumers, and entrepreneurs. Although they belong to the same broad group of fungi, they differ significantly in habitat, cultivation, availability, and commercial importance."

Learning Objectives

After completing this chapter, you will be able to:

- Differentiate between *Cordyceps sinensis* and *Cordyceps militaris.*
- Understand why *C. militaris* is preferred for commercial cultivation.
- Compare their biological characteristics and habitats.
- Evaluate their medicinal and commercial significance.

- Make informed decisions regarding cultivation and product selection.

5.1 Introduction

When people hear the word **Cordyceps**, they often think of the famous "Himalayan Caterpillar Fungus." While this is understandable, it represents only one member of a much larger group of fungi.

Among the hundreds of identified species, **Cordyceps sinensis** (now scientifically classified as *Ophiocordyceps sinensis*) and **Cordyceps militaris** are the two most widely recognized. Both have attracted scientific and commercial interest, but they differ greatly in origin, biology, cultivation, and availability.

For centuries, wild *Cordyceps sinensis* was considered the "gold standard" because it was collected from the Himalayan plateau and was extremely rare. However, advances in biotechnology have enabled the successful cultivation of *Cordyceps militaris*, making it the preferred species for research and commercial production.

This chapter explores these two remarkable fungi in detail.

5.2 Scientific Classification

Although both fungi belong to the same broader group, taxonomic revisions based on molecular research have distinguished them.

Characteristic	***Cordyceps militaris***	***Ophiocordyceps sinensis***
Kingdom	Fungi	Fungi
Division	Ascomycota	Ascomycota
Order	Hypocreales	Hypocreales
Family	Cordycipitaceae	Ophiocordycipitace
Main Habitat	Artificial and natural	High Himalayan regions
Commercial Cultivation	Widely practiced	Extremely difficult

Modern DNA studies have shown that *Ophiocordyceps sinensis* belongs to a different genus, although the older name "Cordyceps sinensis" is still commonly used in trade and popular literature.

5.3 Natural Habitat

Ophiocordyceps sinensis

This species grows naturally at elevations of **3,000–5,500 metres** in alpine meadows.

Its distribution includes:

- Tibet
- Nepal
- Bhutan
- Northern India
- Western China

The harsh climate, short growing season, and dependence on specific insect hosts make natural populations highly limited.

Cordyceps militaris

This species occurs naturally in temperate forests but adapts exceptionally well to artificial cultivation.

It can be grown successfully in:

- Research laboratories
- Mushroom farms
- Universities
- Commercial production units
- Small home laboratories

This adaptability has greatly expanded its availability worldwide.

5.4 Host Relationship

One of the most important differences lies in the relationship with insect hosts.

Wild *Ophiocordyceps sinensis*

Naturally infects the larvae of ghost moths (*Thitarodes* species). The fungus colonizes the larva underground before producing a fruiting body that emerges above the soil.

This highly specialized relationship makes artificial cultivation extremely challenging.

Cordyceps militaris

In nature, it may infect various insects. However, under laboratory conditions, it can be cultivated on nutrient-rich artificial substrates without requiring live insects.

This characteristic makes large-scale production practical and sustainable.

5.5 Appearance

Wild *Ophiocordyceps sinensis*

- Long, slender fruiting body
- Dark brown to black colour
- Attached to a naturally mummified caterpillar
- Irregular shape
- Considerable variation in size

Cordyceps militaris

- Bright orange to deep orange
- Club-shaped fruiting bodies
- Uniform appearance
- Attractive colour

- Suitable for commercial packaging

Its vibrant orange colour is one of the reasons it is popular in health products and educational demonstrations.

5.6 Cultivation

This is where the greatest difference exists.

Ophiocordyceps sinensis

Commercial cultivation remains highly complex because it depends on interactions with specific insect hosts and environmental conditions that are difficult to replicate.

Although researchers continue to investigate cultivation methods, consistent large-scale production remains challenging.

Cordyceps militaris

Can be cultivated using:

- Brown rice
- Silkworm pupae powder
- Nutrient-enriched media

- Liquid culture
- Tissue culture techniques

Controlled environmental conditions make production reliable and repeatable.

5.7 Market Availability

Wild Species

Supply is limited.

Collection depends entirely on natural harvests.

Availability varies from year to year.

Cultivated Species

Available throughout the year.

Can be produced in commercial quantities.

Quality can be standardized through controlled cultivation.

5.8 Price Comparison

Wild *Ophiocordyceps sinensis* is among the most expensive medicinal fungi in the world due to its rarity and labour-intensive collection. Cultivated *Cordyceps militaris* is considerably more affordable because it can be produced efficiently under controlled conditions. This lower production cost has made Cordyceps-based products accessible to a much wider range of consumers.

5.9 Sustainability

Overharvesting of wild populations has raised concerns about environmental conservation and the long-term survival of natural habitats. Cultivated *Cordyceps militaris* offers several sustainability advantages:

- Reduced pressure on wild ecosystems
- Year-round production
- Consistent quality
- Lower environmental impact
- Better traceability

Responsible cultivation also supports research without relying on limited natural resources.

5.10 Research and Commercial Importance

Today, *Cordyceps militaris* has become the primary focus of cultivation research because it is easier to grow and provides a dependable source of material for scientific studies. Its applications include:

- Functional foods
- Nutritional supplements
- Biotechnology research
- Academic teaching
- Commercial mushroom farming
- Product development

Researchers continue to investigate its bioactive compounds and cultivation methods to improve quality and production efficiency.

5.11 Which Species Should Beginners Grow?

For beginners, hobbyists, students, and entrepreneurs, **Cordyceps militaris** is the recommended choice because:

- It adapts well to artificial cultivation.

- It does not require specialised high-altitude conditions.
- Standard cultivation techniques are well established.
- It is suitable for laboratory learning and commercial production.
- It offers consistent results when proper cultivation practices are followed.

Attempting to cultivate wild *Ophiocordyceps sinensis* is generally unsuitable for beginners due to its biological complexity.

Practical Tips

- Purchase cultures only from reputable laboratories or recognised culture collections.
- Keep detailed cultivation records to improve consistency.
- Avoid purchasing products that make unrealistic health claims.
- Learn the difference between wild-collected and cultivated products before

starting a business.
- Focus on quality and hygiene rather than production speed.

Common Misconceptions

Myth: Wild Cordyceps is always superior in every respect.

Fact: Wild and cultivated species differ biologically. Product quality depends on multiple factors, including species, cultivation methods, processing, and quality control. Scientific evaluation should be based on evidence rather than assumptions.

Chapter Summary

In this chapter, you learned:

- The scientific differences between *Ophiocordyceps sinensis* and *Cordyceps militaris*.
- Their natural habitats and ecological adaptations.
- Differences in cultivation, appearance,

availability, and commercial value.

- Why *Cordyceps militaris* has become the preferred species for research and cultivation.
- The importance of sustainability in protecting natural Cordyceps populations.

Key Takeaways

✓ *Cordyceps militaris* is the preferred species for cultivation and commercial production.

✓ Wild *Ophiocordyceps sinensis* remains rare and difficult to cultivate.

✓ Sustainable cultivation helps protect fragile Himalayan ecosystems.

✓ Modern biotechnology has made high-quality *Cordyceps militaris* available worldwide.

Chapter 6 : Bioactive Compounds of Cordyceps militaris

The Science Behind Its Medicinal Value

"The remarkable reputation of Cordyceps is not due to a single miracle compound but to a complex mixture of naturally occurring bioactive substances that work together within the fungus."

Learning Objectives

After completing this chapter, you will be able to:

- Understand what bioactive compounds are.
- Identify the major compounds found in *Cordyceps militaris.*
- Learn their biological functions.
- Understand why cultivation conditions influence chemical composition.
- Appreciate the importance of quality control in producing high-quality Cordyceps.

6.1 Introduction

Medicinal fungi have attracted worldwide scientific attention because they produce a diverse range of naturally occurring chemical compounds. Unlike ordinary food ingredients that primarily provide energy and nutrition, these compounds may influence various biological processes within living organisms.

The medicinal value of *Cordyceps militaris* is largely attributed to its **bioactive compounds**. These molecules are synthesized naturally during the growth and development of the fungus and contribute to its unique biological characteristics.

It is important to recognize that no single compound can explain all of the biological activities associated with Cordyceps. Instead, researchers believe that multiple compounds act together, creating a complex biochemical profile. The concentration of these compounds varies depending on:

- Genetic strain

- Culture quality
- Nutrient composition
- Temperature
- Humidity
- Light exposure
- Harvest stage
- Drying method
- Storage conditions

For this reason, proper cultivation techniques are essential for producing high-quality fruiting bodies.

6.2 What Are Bioactive Compounds?

A **bioactive compound** is a naturally occurring chemical substance that can interact with biological systems. Unlike carbohydrates, proteins, and fats, which mainly provide nutrition, bioactive compounds often perform specialized functions within living organisms. In fungi, these compounds may help:

- Protect against environmental stress

- Resist microbial competitors
- Regulate growth
- Facilitate reproduction
- Adapt to changing environments

Scientists study these molecules to understand their properties and evaluate their potential applications in food science, biotechnology, and medicine.

6.3 Major Bioactive Compounds Found in *Cordyceps militaris*

Research has identified numerous naturally occurring compounds in *Cordyceps militaris*. The most important include:

- Cordycepin
- Adenosine
- Polysaccharides
- Beta-glucans
- Ergosterol
- D-Mannitol
- Proteins
- Essential amino acids

- Peptides
- Polyphenols
- Flavonoid-like compounds
- Carotenoid pigments
- Vitamins
- Minerals

Each contributes differently to the biological profile of the fungus.

6.4 Cordycepin

Cordycepin (3'-deoxyadenosine) is considered one of the signature compounds of *Cordyceps militaris.* Its chemical structure closely resembles the naturally occurring nucleoside adenosine but differs by the absence of one oxygen atom. This seemingly small structural difference gives cordycepin unique biological properties that have attracted significant research interest.

Why Is Cordycepin Important?

Cordycepin has been investigated in laboratory and animal studies for its potential roles in:

- Cellular regulation
- Inflammatory pathways
- Oxidative stress responses
- Microbial interactions
- Cell signalling

Research is ongoing, and while findings are promising, further well-designed human clinical studies are needed before definitive medical conclusions can be drawn.

Factors Affecting Cordycepin Content

The concentration of cordycepin depends on:

- Genetic strain
- Substrate composition
- Temperature
- Incubation period
- Light intensity
- Harvest timing

For commercial cultivators, optimizing these factors is essential to maximize product quality.

6.5 Adenosine

Adenosine is another naturally occurring nucleoside found in *Cordyceps militaris*. Within living cells, adenosine plays important roles in energy transfer and cellular communication. It is a normal component of many biological systems. In Cordyceps, adenosine contributes to the fungus's overall chemical profile and is commonly measured during quality evaluation. Researchers continue to investigate its significance in conjunction with other bioactive compounds.

6.6 Polysaccharides

Polysaccharides are long chains of sugar molecules naturally synthesized by fungi. They serve several important functions:

- Energy storage
- Cell wall structure
- Environmental protection
- Growth regulation

Fungal polysaccharides have attracted considerable scientific interest because of their interactions with biological systems. The composition and quantity of polysaccharides vary depending on cultivation conditions and fungal strain.

6.7 Beta-Glucans

Beta-glucans are specialized polysaccharides found in the cell walls of many fungi. These compounds have become important subjects of nutritional and immunological research. Healthy fruiting bodies generally contain higher levels of structurally intact beta-glucans than poorly developed or contaminated cultures. Proper drying and storage help preserve these compounds.

6.8 Ergosterol

Ergosterol is the principal sterol found in fungal cell membranes. It performs functions similar to cholesterol in animal cells by maintaining membrane stability and flexibility. Exposure to ultraviolet light can convert ergosterol into vitamin D_2, making mushrooms one of the few natural non-animal sources of this vitamin. Ergosterol content is often used as an indicator of fungal biomass and quality.

6.9 D-Mannitol

D-Mannitol, sometimes called cordycepic acid in older literature (though the terms are not strictly equivalent), is a naturally occurring sugar alcohol present in Cordyceps. It contributes to osmotic regulation within fungal cells and has attracted scientific interest because of its biochemical properties.

6.10 Proteins and Amino Acids

Like other edible fungi, *Cordyceps militaris* contains proteins composed of amino acids. Essential amino acids identified in cultivated fruiting bodies include:

- Leucine
- Isoleucine
- Valine
- Lysine

- Methionine
- Phenylalanine
- Threonine

Protein content varies according to substrate composition and cultivation conditions.

6.11 Vitamins and Minerals

Cordyceps contains small amounts of vitamins and minerals that contribute to its nutritional value. Commonly reported nutrients include:

Vitamins

- B_1 (Thiamine)
- B_2 (Riboflavin)
- B_3 (Niacin)
- B_5 (Pantothenic acid)

Minerals

- Potassium
- Phosphorus
- Magnesium

- Calcium
- Iron
- Zinc
- Selenium (trace amounts)

The exact composition depends on the growing medium and production methods.

6.12 Pigments

The bright orange colour of *Cordyceps militaris* results from naturally occurring carotenoid pigments. Healthy pigmentation often indicates:

- Proper lighting
- Good nutrition
- Appropriate environmental conditions

However, colour alone should not be used as the sole indicator of quality.

6.13 Factors Affecting Bioactive Compound Production

Cultivation practices strongly influence chemical composition.

Important factors include:

1. Genetics

Different strains naturally produce different quantities of bioactive compounds.

2. Nutrient Medium

Balanced nutrition encourages healthy development and compound synthesis.

3. Temperature

Improper temperatures can reduce growth and alter chemical composition.

4. Light

Controlled light exposure influences pigmentation and fruiting.

5. Oxygen

Adequate air exchange supports healthy metabolism.

6. Harvest Timing

Harvesting too early or too late may affect the concentration of certain compounds.

7. Drying

Gentle drying at appropriate temperatures helps preserve heat-sensitive components.

6.14 Quality Evaluation

Commercial producers often evaluate their products using laboratory techniques such as:

- High-Performance Liquid Chromatography (HPLC)
- Moisture analysis
- Microbial testing
- Heavy metal testing
- DNA authentication
- Microscopic examination

These methods help ensure product quality, safety, and consistency.

Practical Tips for Growers

✔ Choose genetically stable, high-quality strains.

✔ Maintain clean cultivation conditions.

✔ Record temperature and humidity daily.

✔ Harvest at the recommended maturity stage.

✔ Dry fruiting bodies carefully to preserve quality.

✔ Store dried products in airtight, moisture-resistant containers away from direct sunlight.

Common Mistakes

⟐ Assuming all orange fruiting bodies contain the same level of bioactive compounds.

⟐ Using old or genetically unstable cultures.

⟐ Overheating during drying.

⟐ Storing dried products in humid environments.

⟐ Ignoring quality testing.

Chapter Summary

In this chapter, you learned:

- What bioactive compounds are.
- The major naturally occurring compounds in *Cordyceps militaris*.
- The importance of cordycepin, adenosine, polysaccharides, beta-glucans, ergosterol, proteins, vitamins, and minerals.
- How cultivation conditions influence chemical composition.
- Why quality control and standardized cultivation practices are essential.

Understanding these compounds provides the scientific basis for appreciating the potential value of *Cordyceps militaris*. In the next chapter, we will examine **how these compounds interact with the human body**, exploring current scientific understanding of their roles in metabolism, immunity, oxidative stress, and other biological processes, while clearly distinguishing established evidence from ongoing research.

Chapter 7 : How Cordyceps Works in the Human Body

Biological Mechanisms and Scientific Perspectives

"Understanding how Cordyceps interacts with the human body helps readers appreciate both its potential and its limitations. Scientific knowledge allows us to distinguish between evidence-based information and unsupported claims."

Learning Objectives

After studying this chapter, you will be able to:

- Understand how nutrients and bioactive compounds from *Cordyceps militaris* are processed in the body.
- Learn about the major biological systems that researchers are investigating.
- Understand the concepts of oxidative stress, inflammation, and immune regulation.
- Recognize the difference between laboratory findings, animal studies, and

human clinical evidence.

- Appreciate the importance of responsible and evidence-based use.

7.1 Introduction

The human body is a highly organized biological system in which billions of cells communicate continuously to maintain health. Every second, chemical reactions occur that regulate growth, repair damaged tissues, produce energy, fight infections, and maintain internal balance.

Foods and natural products influence these processes by providing nutrients and biologically active compounds. Once consumed, these compounds are digested, absorbed, transported through the bloodstream, and eventually reach different tissues and organs.

Cordyceps militaris contains several naturally occurring compounds that have attracted scientific interest because they interact with various biological pathways. Researchers continue to study these interactions to understand their significance and potential applications.

It is important to remember that biological activity does not automatically mean therapeutic effectiveness. The human body is complex, and results observed in laboratory experiments may not always translate directly to clinical outcomes. Therefore, scientific conclusions should be based on carefully conducted research.

7.2 Digestion and Absorption

After consumption, dried Cordyceps powder, capsules, extracts, or food products enter the digestive system. The digestive process involves several stages:

Mouth

Chewing breaks the material into smaller particles, increasing the surface area for digestion.

Stomach

Acids and digestive enzymes begin breaking down proteins and other components.

Small Intestine

Most nutrients and many bioactive compounds are absorbed through the intestinal wall into the bloodstream.

Liver

The absorbed compounds are transported to the liver, where they undergo various metabolic processes before being distributed throughout the body.

Distribution

The bloodstream carries absorbed compounds to different organs, where they may participate in normal physiological processes.

7.3 The Immune System

One of the most widely studied areas of Cordyceps research is its interaction with the immune system. The immune system consists of specialized cells, tissues, and organs that protect the body from harmful microorganisms while maintaining tolerance toward normal tissues.

Major immune components include:

- White blood cells
- Macrophages
- Neutrophils
- Natural killer (NK) cells
- T lymphocytes
- B lymphocytes
- Cytokines

Laboratory research suggests that certain fungal polysaccharides may interact with immune cells, but the exact effects in humans depend on many factors and continue to be investigated.

7.4 Oxidative Stress

Every cell in the body naturally produces reactive molecules during normal metabolism. These molecules are known as **reactive oxygen species (ROS)**.In moderate amounts, ROS are essential for normal cellular functions. However, excessive accumulation may contribute to oxidative stress. Oxidative stress has been associated with aging and various chronic diseases.

Many foods, including fruits, vegetables, and mushrooms, contain antioxidant compounds that help maintain the body's natural balance.

Researchers are studying the antioxidant properties of compounds found in *Cordyceps militaris*, although the significance of these findings in human health requires further investigation.

7.5 Inflammation

Inflammation is a natural protective response that helps the body repair damaged tissues and defend against infections. There are two major types:

Acute Inflammation

A short-term response that usually resolves after healing. Examples include:

- Cuts
- Minor burns
- Infections

Chronic Inflammation

Long-term inflammation may occur when the normal healing process is disrupted. Researchers have investigated whether compounds from Cordyceps influence inflammatory pathways in laboratory models, but additional clinical research is needed to determine their role in human health.

7.6 Cellular Energy

Every living cell requires energy to function. This energy is primarily produced inside microscopic structures called **mitochondria**. The energy molecule **ATP (adenosine triphosphate)** powers countless cellular activities. Because Cordyceps contains adenosine-related compounds, scientists have investigated its possible relationship with cellular energy metabolism. Some laboratory and animal studies have reported interesting findings, particularly in exercise physiology, but human evidence remains an active area of research.

7.7 Respiratory Function

Historically, Cordyceps has been associated with respiratory wellness in traditional medicine. Modern researchers have explored this area by examining:

- Oxygen utilization
- Exercise capacity
- Respiratory physiology
- Cellular metabolism

While some studies suggest potential benefits under specific conditions, results are not always consistent, and further high-quality clinical trials are required.

7.8 Metabolism

Metabolism refers to all chemical reactions occurring within the body.

These reactions include:

- Digestion
- Energy production
- Protein synthesis

- Fat metabolism
- Carbohydrate metabolism
- Hormone regulation

Scientists continue investigating how fungal compounds interact with metabolic pathways, but current evidence should be interpreted cautiously.

7.9 Scientific Research: Levels of Evidence

Understanding the quality of evidence is essential. Not all research carries equal weight.

Laboratory Studies (In Vitro)

Experiments performed using isolated cells or microorganisms. These studies help identify possible biological mechanisms but do not demonstrate clinical effectiveness.

Animal Studies

Research conducted using laboratory animals. Animal experiments provide valuable information but cannot fully predict human responses.

Human Clinical Trials

Studies involving human volunteers. Well-designed randomized controlled trials provide stronger evidence than laboratory or animal studies.

Systematic Reviews and Meta-Analyses

These combine data from multiple clinical studies and generally provide the highest level of scientific evidence. Throughout this book, health-related discussions will distinguish between these different levels of evidence.

7.10 Responsible Use

Consumers should remember several important principles.

- Cordyceps is **not** a substitute for a balanced diet.
- It should **not** replace prescribed medications unless advised by a qualified healthcare professional.
- Individuals with chronic illnesses should consult their healthcare provider before

using dietary supplements.

- Product quality varies considerably among manufacturers.
- Choosing products from reputable sources is essential.

Practical Tips

✔ Read product labels carefully.

✔ Store dried Cordyceps in a cool, dry place.

✔ Follow recommended serving instructions.

✔ Maintain a healthy lifestyle that includes balanced nutrition, regular physical activity, adequate sleep, and stress management.

✔ View Cordyceps as one component of overall wellness rather than a standalone solution.

Common Misconceptions

Myth: If a compound shows activity in a laboratory experiment, it will definitely cure diseases in humans.

Fact: Laboratory findings are an important first step, but they must be confirmed through carefully designed clinical trials before medical conclusions can be made.

Myth: More Cordyceps always produces better results.

Fact: Excessive consumption of any supplement is not necessarily beneficial. Responsible use and adherence to recommended guidance are important.

Chapter Summary

In this chapter, you learned:

- How bioactive compounds are absorbed and distributed in the human body.
- The basic functions of the immune system.
- The concepts of oxidative stress and inflammation.
- How cells produce energy.
- The importance of evaluating scientific evidence according to its quality.
- Why responsible use and realistic expectations are essential.

Key Takeaways

✓ The human body is highly complex, and natural products interact with many biological systems.

✓ Laboratory research provides valuable insights but does not always predict clinical outcomes.

✓ Understanding the levels of scientific evidence helps readers make informed decisions.

✓ High-quality cultivation and processing are essential for producing consistent Cordyceps products.

Chapter 8 : Evidence-Based Health Benefits of Cordyceps militaris

"Cordyceps has been used in traditional medicine for centuries. Modern science seeks to understand which of these traditional claims are supported by evidence, which remain uncertain, and where further research is needed."

Learning Objectives

After studying this chapter, you will be able to:

- Understand the traditional uses of *Cordyceps militaris*.
- Evaluate current scientific evidence regarding its potential health benefits.
- Distinguish between laboratory studies, animal research, and human clinical trials.
- Develop realistic expectations about the role of Cordyceps in health and wellness.
- Understand why quality, dosage, and lifestyle influence outcomes.

8.1 Introduction

The popularity of *Cordyceps militaris* has increased dramatically over the past two decades. Health-conscious consumers, athletes, researchers, and healthcare professionals have shown growing interest in its unique bioactive compounds and their potential biological activities.

Traditional medical systems have long described Cordyceps as a natural tonic that promotes vitality, endurance, and overall well-being. Modern research has investigated these claims using laboratory experiments, animal models, and human studies. While many findings are promising, not all traditional claims have been confirmed, and research is still evolving.

This chapter reviews the current scientific understanding of the potential health benefits of *Cordyceps militaris*, presenting the evidence in a balanced and practical manner.

8.2 Immune System Support

Traditional Perspective

For centuries, Cordyceps has been used to promote general health and resilience, especially during recovery from illness and periods of physical weakness.

Scientific Understanding

Research suggests that polysaccharides and beta-glucans present in Cordyceps may interact with immune cells such as macrophages, dendritic cells, and natural killer (NK) cells. These interactions are being studied for their potential role in supporting normal immune function. It is important to note that "immune support" does not mean simply making the immune system stronger. A healthy immune system functions through balanced regulation, responding appropriately to threats while avoiding excessive or inappropriate activation.

Strength of Evidence

- Laboratory studies: Strong
- Animal studies: Moderate to strong
- Human studies: Limited but growing

Practical Note

Cordyceps should be viewed as a dietary supplement that may support overall wellness. It should not be considered a treatment or preventive measure for infectious diseases.

8.3 Physical Performance and Exercise

One of the most widely publicized uses of Cordyceps is its potential to support physical endurance.

Historical Background

Traditional healers believed that Cordyceps increased stamina and reduced fatigue. Interest in this area expanded after reports of improved athletic performance among some Chinese athletes in the 1990s, although the exact role of Cordyceps in those performances remains uncertain.

Current Research

Scientists have investigated whether Cordyceps influences:

- Oxygen utilization
- Exercise capacity
- Recovery after physical activity
- Cellular energy metabolism

Some studies suggest modest improvements in exercise performance, particularly in older adults or individuals with lower baseline fitness. Results in highly trained athletes have been mixed.

Strength of Evidence

- Laboratory studies: Moderate
- Human studies: Mixed

Practical Advice

Cordyceps should not be viewed as a performance-enhancing drug. Regular exercise, balanced nutrition, adequate hydration, and proper recovery remain the cornerstones of athletic performance.

8.4 Fatigue and Energy

Fatigue is one of the most common reasons people seek natural health products. Researchers have explored whether Cordyceps may influence factors related to energy metabolism and recovery. Several laboratory and animal studies have reported encouraging findings, but human evidence is still limited and inconsistent. Fatigue has many possible causes, including:

- Poor sleep
- Stress
- Nutritional deficiencies
- Medical conditions
- Certain medications

Individuals experiencing persistent fatigue should seek medical evaluation rather than relying solely on dietary supplements.

8.5 Respiratory Health

Cordyceps has traditionally been associated with respiratory wellness.

Researchers have investigated its potential influence on:

- Lung function
- Oxygen utilization
- Exercise-related breathing
- Recovery following respiratory stress

Although some studies report encouraging observations, additional high-quality clinical trials are needed before firm conclusions can be drawn. Maintaining respiratory health also depends on avoiding smoking, minimizing air pollution exposure, staying physically active, and following recommended vaccinations where appropriate.

8.6 Healthy Aging

Aging is associated with gradual changes in cellular function, metabolism, and tissue repair. Scientists are investigating whether the antioxidant compounds found in Cordyceps help protect cells from oxidative stress, one of many factors involved in the aging process. Current evidence suggests that Cordyceps may contribute to overall wellness as part of a healthy lifestyle, but it should not be regarded as an anti-aging cure. Healthy aging depends on multiple factors, including:

- Nutritious diet
- Regular exercise
- Quality sleep
- Stress management
- Social engagement
- Preventive healthcare

8.7 Cognitive Function

Researchers have also examined whether Cordyceps influences cognitive health. Studies have explored possible effects on:

- Learning
- Memory
- Mental alertness
- Neuroprotection

Most findings currently come from laboratory and animal research.

Human evidence remains limited, and more well-designed clinical studies are required before recommendations can be made.

8.8 Cardiovascular Health

The cardiovascular system includes the heart and blood vessels. Researchers continue investigating how Cordyceps compounds may influence:

- Blood circulation
- Oxidative stress

- Lipid metabolism
- Cellular health

Some experimental studies have produced promising results, but further human research is required. Individuals with heart disease should always follow the advice of their healthcare provider and should not discontinue prescribed medications.

8.9 Metabolic Health

Metabolism refers to the body's processes for converting food into energy and maintaining normal physiological functions.

Scientists have explored the possible relationship between Cordyceps and:

- Glucose metabolism
- Lipid metabolism
- Body weight regulation

Although early research is encouraging, existing evidence is insufficient to recommend Cordyceps as a treatment for metabolic diseases.

Lifestyle interventions remain the most effective approach to maintaining metabolic health.

8.10 Liver and Kidney Health

The liver and kidneys perform essential functions, including detoxification, metabolism, and waste removal. Experimental studies have investigated the effects of Cordyceps compounds on these organs. While laboratory findings are interesting, they do not justify replacing established medical care. People with liver or kidney disease should consult their healthcare provider before using any dietary supplement.

8.11 Reproductive Health

Traditional medicine has long associated Cordyceps with reproductive vitality. Modern researchers have explored this topic in laboratory and animal studies. Although some findings suggest possible biological activity, human clinical evidence remains limited. Therefore, no medical claims regarding fertility or reproductive disorders should be made without robust scientific evidence.

8.12 Antioxidant Activity

One of the best-established properties of many edible and medicinal mushrooms is their antioxidant content. Antioxidants help neutralize excess reactive oxygen species produced during normal metabolism. A diet rich in fruits, vegetables, whole grains, legumes, nuts, and mushrooms—including *Cordyceps militaris*—may contribute to a balanced intake of naturally occurring antioxidant compounds.

8.13 Cancer Research: A Balanced Perspective

Cancer research involving Cordyceps has received considerable attention. Scientists have investigated cordycepin and other compounds in:

- Cell culture studies
- Animal experiments
- Mechanistic research

These studies have provided valuable insights into cellular biology. However, laboratory findings should not be interpreted as proof that Cordyceps prevents or cures cancer in humans. Patients diagnosed with cancer should always follow evidence-based medical treatment recommended by qualified oncology specialists. Cordyceps may be an interesting area of future research, but more clinical evidence is needed.

Practical Tips

✔ Purchase products from reputable manufacturers.

✔ Follow the recommended serving size.

✔ Maintain realistic expectations.

✔ Consult a healthcare professional if you have chronic medical conditions or take prescription medications.

✔ Remember that dietary supplements complement—but do not replace—a healthy lifestyle.

Myth vs. Fact

Myth	Fact
Cordyceps cures every disease.	No scientific evidence supports such a claim.
More Cordyceps always produces greater benefits.	Excessive intake is not necessarily better. Follow recommended guidance.
Natural products are always completely safe.	Natural products can interact with medications or medical conditions.
One research study proves effectiveness.	Scientific understanding develops through multiple well-designed studies.

Chapter Summary

In this chapter, you have learned:

- The traditional uses of *Cordyceps militaris*.
- Current scientific evidence regarding immune function, exercise, fatigue, respiratory health, healthy aging, cognition, cardiovascular health, metabolism, liver and kidney health, and antioxidant activity.

- Why the strength of evidence varies across different health topics.
- The importance of interpreting research responsibly.
- How Cordyceps fits within an overall healthy lifestyle rather than serving as a replacement for medical care.

Key Takeaways

✓ *Cordyceps militaris* contains biologically active compounds that are being actively studied.

✓ Some research areas show promising results, while others require additional clinical evidence.

✓ Consumers should rely on evidence-based information rather than exaggerated marketing claims.

✓ Responsible use, product quality, and a healthy lifestyle are essential for obtaining the greatest potential value from Cordyceps.

Chapter 9: Safety, Dosage, Side Effects, Storage, and Quality Assurance of Cordyceps militaris

"The true value of a medicinal mushroom lies not only in its cultivation but also in its safe, responsible, and scientifically informed use."

Learning Objectives

After completing this chapter, you will be able to:

- Understand the safety profile of *Cordyceps militaris*.
- Learn appropriate dosage principles.
- Identify possible side effects and precautions.
- Recognize interactions with medicines.
- Select high-quality Cordyceps products.
- Store Cordyceps correctly to preserve its quality.
- Understand quality assurance and regulatory considerations.

9.1 Introduction

The increasing popularity of *Cordyceps militaris* has resulted in a rapidly expanding global market. Today, consumers can purchase dried fruiting bodies, powders, capsules, tablets, extracts, teas, coffees, chocolates, and functional beverages containing Cordyceps. While this increased availability offers many choices, it also creates important questions.

- Which products are genuine?
- How much should be consumed?
- Are there any side effects?
- Can it interact with medicines?
- How should it be stored?
- How can consumers identify premium-quality products?

This chapter provides practical answers based on current scientific knowledge and accepted food safety principles.

9.2 Is *Cordyceps militaris* Safe?

When cultivated under hygienic conditions and consumed responsibly as a food or dietary supplement, *Cordyceps militaris* is generally considered to have a favorable safety profile. However, safety depends on several factors:

- Product quality
- Correct species identification
- Cultivation hygiene
- Freedom from contaminants
- Appropriate consumption
- Individual health status

A high-quality product produced under controlled laboratory conditions is far safer than material collected from unknown or contaminated sources.

9.3 Who Should Consult a Healthcare Professional Before Use?

Although Cordyceps is widely consumed, certain individuals should seek professional medical advice before using it regularly. These include:

Pregnant Women

Research regarding safety during pregnancy remains limited. Therefore, routine supplementation should only be considered under medical supervision.

Breastfeeding Mothers

Scientific information is currently insufficient. Professional guidance is recommended.

Children

Children have different nutritional and physiological requirements. Parents should consult qualified healthcare professionals before introducing any dietary supplement.

People with Autoimmune Diseases

Because researchers continue investigating how Cordyceps interacts with immune function, individuals diagnosed with autoimmune disorders should seek medical advice before use.

Organ Transplant Recipients

People taking immunosuppressive medications should consult their transplant specialist before using any herbal or fungal supplements.

Individuals with Serious Chronic Illness

Patients receiving treatment for conditions such as:

- cancer
- liver disease
- kidney disease
- cardiovascular disease
- diabetes

should discuss supplement use with their healthcare provider to avoid possible interactions or misunderstandings regarding their treatment plan.

9.4 Possible Side Effects

Most healthy individuals tolerate *Cordyceps militaris* well. However, mild side effects have occasionally been reported. These may include:

- mild stomach discomfort
- nausea
- bloating
- dry mouth
- digestive upset

These symptoms are generally temporary and often resolve after reducing intake or discontinuing use. Severe allergic reactions are uncommon but require immediate medical attention.

9.5 Allergic Reactions

Although rare, allergy to mushrooms or fungal products is possible.

Symptoms may include:

- skin rash
- itching
- swelling
- difficulty breathing

Individuals with known mushroom allergies should exercise caution and seek medical advice before consuming Cordyceps products.

9.6 Can Cordyceps Interact with Medicines?

Researchers continue studying possible interactions. Because evidence remains limited, caution is advisable. Individuals taking medications such as:

- blood-thinning medicines
- medicines affecting immune function
- diabetes medicines
- medicines used after organ transplantation

should consult their healthcare provider before introducing any new supplement. Never stop prescribed medication without medical advice.

9.7 Dosage Principles

One of the most frequently asked questions is: **"How much Cordyceps should I take?"** There is **no universally accepted dosage** suitable for everyone. Appropriate intake depends upon:

- age
- health condition
- product type
- concentration
- manufacturing process
- advice provided by healthcare professionals

Commercial products should always be used according to the manufacturer's instructions unless otherwise advised by a qualified practitioner. For research purposes, different studies have used varying amounts, making direct comparisons difficult. Consumers should avoid assuming that higher quantities automatically provide greater benefits.

9.8 Importance of Product Standardization

Two products labelled "Cordyceps" may differ greatly in quality. Factors influencing quality include:

- fungal strain
- cultivation method
- substrate
- harvesting stage
- drying technique
- extraction process
- storage conditions

Standardized production helps ensure consistency between batches.

Commercial producers should maintain detailed production records and quality control procedures.

9.9 How to Identify a High-Quality Product

Consumers should examine several characteristics before purchasing.

Correct Scientific Name

The label should clearly state:

Cordyceps militaris

Scientific names help distinguish products from unrelated fungi.

Manufacturer Information

Choose products from reputable manufacturers who provide:

- complete contact information
- manufacturing details
- batch number
- expiry date

Packaging

Good packaging protects the product from:

- moisture
- sunlight
- oxygen
- contamination

Vacuum-sealed or moisture-resistant packaging is generally preferable.

Laboratory Testing

Reliable manufacturers may provide information regarding:

- microbial testing
- heavy metal testing
- pesticide testing
- identity verification
- moisture content

Third-party laboratory testing increases consumer confidence.

9.10 Storage Guidelines

Proper storage preserves quality.

Fresh Fruiting Bodies

Store in a refrigerator.

Consume promptly according to food safety recommendations.

Dried Fruiting Bodies

Store in:

- airtight containers
- cool environment
- dry location
- away from direct sunlight

Moisture is the greatest enemy of dried mushrooms.

Powder

Keep tightly sealed after every use.

Avoid introducing moisture by using dry utensils.

Extracts

Follow the manufacturer's storage recommendations.

9.11 Shelf Life

Shelf life depends upon:

- drying quality

- moisture content
- packaging
- storage conditions

Properly dried and sealed products generally remain stable longer than products exposed to humidity. However, consumers should always follow expiry dates provided by the manufacturer.

9.12 Signs of Product Deterioration

Discard the product if you observe:

- mould growth
- unusual odour
- insect infestation
- excessive moisture
- abnormal colour changes
- damaged packaging

Food safety should always take priority over cost.

9.13 Quality Assurance in Commercial Production

Commercial cultivators should implement Good Manufacturing Practices (GMP).

Essential quality control includes:

Raw Material Inspection

Only high-quality ingredients should be accepted.

Sterility Monitoring

Laboratory hygiene must be maintained throughout production.

Environmental Monitoring

Temperature, humidity, lighting, and air quality should be recorded daily.

Batch Documentation

Each production batch should receive:

- batch number
- production date
- substrate details

- culture identification
- harvest date

Good documentation allows traceability.

Product Testing

Finished products should be evaluated for:

- moisture
- microbial contamination
- appearance
- colour
- physical integrity

Where feasible, producers may also analyse important compounds such as cordycepin using validated laboratory methods.

9.14 Regulatory Considerations

Regulations vary between countries.

Commercial producers should comply with all applicable requirements related to:

- food safety

- labelling
- packaging
- hygiene
- quality standards
- licensing

Products should never make unsupported medical claims.

Health information should accurately reflect the available scientific evidence.

Practical Tips

✓ Purchase products only from trusted manufacturers.

✓ Read labels carefully before use.

✓ Store products in cool, dry conditions.

✓ Keep containers tightly sealed.

✓ Record opening dates for long-term storage.

✓ Follow recommended usage instructions.

✓ Seek professional advice if you have chronic medical conditions or take prescription medicines.

Frequently Asked Questions

Can I consume Cordyceps every day?

Many people include Cordyceps as part of their regular diet or supplement routine. However, the appropriate frequency depends on the individual, the product used, and professional guidance where necessary.

Can I cook Cordyceps?

Yes. Cultivated *Cordyceps militaris* can be incorporated into soups, herbal teas, porridges, and other culinary preparations. Excessive heating, however, may affect some heat-sensitive compounds, so gentle cooking is generally preferred.

Can I store dried Cordyceps in the refrigerator?

If the product is thoroughly dried and protected from moisture, refrigeration is not usually necessary. A cool, dry cupboard is often sufficient. In humid climates, refrigeration in an airtight container may help preserve quality, provided condensation is avoided.

How can I tell whether my dried Cordyceps has spoiled?

Signs such as mould growth, an unpleasant odour, visible moisture, insect activity, or damaged packaging indicate that the product should not be consumed.

Chapter Summary

In this chapter, you learned:

- The general safety profile of *Cordyceps militaris*.
- Who should seek medical advice before use.
- Possible side effects and precautions.
- Principles of responsible consumption.
- How to identify high-quality products.
- Correct storage methods.
- The importance of quality assurance in commercial cultivation.
- The need to comply with food safety and

labelling regulations.

Key Takeaways

✓ Safe cultivation begins with hygienic production and continues through proper storage and responsible use.

✓ Quality matters more than quantity.

✓ Consumers should choose products from reliable manufacturers with transparent quality-control practices.

✓ Evidence-based information and regulatory compliance are essential for building trust in Cordyceps products.

PART II

PRACTICAL CULTIVATION OF CORDYCEPS MILITARIS

Chapter 10: Setting Up a Cordyceps Laboratory: Infrastructure, Equipment, and Sterile Work Practices

"A successful Cordyceps grower is not judged by the number of jars produced but by the cleanliness of the laboratory. Sterility is the foundation upon which every successful cultivation is built."

Learning Objectives

After completing this chapter, you will be able to:

- Design a home-scale or commercial Cordyceps laboratory.
- Understand laboratory zoning and workflow.
- Select essential equipment according to your budget.
- Establish sterile working conditions.
- Prevent contamination through proper laboratory practices.
- Maintain records and laboratory hygiene.

10.1 Introduction

Many beginners believe that cultivating *Cordyceps militaris* requires an expensive laboratory with sophisticated scientific equipment. This misconception discourages aspiring cultivators from entering the field.

In reality, successful cultivation depends far more on **cleanliness, discipline, and proper procedures** than on costly equipment. A small, well-organized laboratory can consistently produce high-quality Cordyceps if sterile techniques are followed carefully. Whether you are a student, hobbyist, researcher, or entrepreneur, your laboratory should be designed to minimize contamination, improve workflow, and ensure consistent environmental control. This chapter explains how to establish an efficient laboratory suitable for both beginners and commercial producers.

10.2 Selecting the Right Location

The first step is choosing an appropriate room. An ideal laboratory should have the following characteristics:

Quiet Environment

Avoid areas with heavy foot traffic. Frequent movement introduces dust and airborne microorganisms.

Dry Environment

Avoid damp rooms. High humidity encourages mould growth.

Good Ventilation

Fresh air is important. However, avoid strong drafts that carry dust into sterile areas.

Easy to Clean

Floors should preferably be:

- tiled
- vinyl
- epoxy coated

Avoid carpets. Walls should have smooth surfaces that can be cleaned regularly.

Adequate Electricity

Ensure reliable power for:

- lighting
- pressure cooker
- autoclave
- incubator
- air purifier
- laminar airflow cabinet

Installing a backup power source is highly recommended for commercial facilities.

10.3 Laboratory Layout

An efficient laboratory should follow a one-way workflow to reduce contamination. The recommended sequence is:

Raw Material Storage → Media Preparation → Sterilization → Inoculation → Incubation → Fruiting Room → Harvesting → Drying → Packaging

Never move contaminated materials back into clean areas.

10.4 Laboratory Zones

For better organization, divide the laboratory into separate functional zones.

Zone 1 – Storage Area

Store:

- rice
- silkworm pupae powder
- bottles
- jars
- chemicals
- packaging materials

Use airtight containers to prevent insect infestation and moisture absorption.

Zone 2 – Washing Area

Used for:

- cleaning bottles

- washing tools
- preparing equipment

Install:

- stainless steel sink
- clean water supply
- drainage system

Zone 3 – Media Preparation Room

This area is used for:

- weighing ingredients
- mixing substrates
- filling containers

Recommended equipment:

- electronic balance
- measuring cylinders
- mixing bowls
- pH meter

Zone 4 – Sterilization Room

Contains:

- autoclave
- pressure cooker
- steam sterilizer

Safety is extremely important.

Never leave sterilization equipment unattended during operation.

Zone 5 – Inoculation Room

This is the most important room in the laboratory. It should remain extremely clean. Recommended equipment:

- Laminar Air Flow Cabinet
- Still Air Box (for beginners)
- UV lamp
- alcohol spray bottle
- gas burner or spirit lamp

Only sterilized materials should enter this room.

Zone 6 – Incubation Room

Purpose: Growth of mycelium. Conditions should remain stable.

Recommended:

Temperature: 20–24°C

Dark or low-light environment

Minimal disturbance

Daily observation

Zone 7 – Fruiting Room

After complete colonization, culture bottles are transferred here.

Maintain:

Temperature

Humidity

Light

Fresh air exchange

CO_2 monitoring (if possible)

Zone 8 – Harvesting and Drying Area

Should remain clean and dust-free.

Activities include:

Harvesting

Cleaning

Drying

Sorting

Packaging

10.5 Essential Laboratory Furniture

A practical laboratory should include:

- Stainless steel working table
- Chemical-resistant shelves
- Lockable storage cabinet
- Plastic storage bins
- Office desk
- Laboratory chair
- Record-keeping cabinet

Avoid wooden furniture in humid environments because it can support mould growth.

10.6 Laboratory Safety

Safety is just as important as production.

Always provide:

First Aid Box

Fire Extinguisher

Emergency Lighting

Protective Gloves

Face Masks

Safety Goggles

Laboratory Apron

Emergency Contact Numbers

Proper Electrical Earthing

10.7 Personal Hygiene

Human beings are the largest source of contamination.

Always:

Wash hands thoroughly.

Wear clean clothing.

Use gloves.

Wear a face mask.

Tie long hair.

Avoid unnecessary talking during inoculation.

Do not eat or drink inside the laboratory.

10.8 Cleaning Schedule

An effective laboratory follows a strict cleaning routine.

Daily

Sweep and mop the floor.

Disinfect working surfaces.

Empty waste bins.

Check incubators.

Inspect cultures.

Weekly

Clean shelves.

Wash windows.

Clean air filters.

Inspect equipment.

Calibrate thermometers.

Monthly

Deep clean laboratory.

Check electrical wiring.

Service equipment.

Review contamination records.

Replace damaged tools.

10.9 Waste Disposal

Proper waste management prevents contamination.

Separate waste into:

Biological Waste

Contaminated cultures

Old substrates

Discarded mushrooms

These should be sterilized before disposal.

Plastic Waste

Bottle caps

Plastic bags

Gloves

Packaging materials

Dispose according to local regulations.

Glass Waste

Broken bottles

Broken Petri dishes

Dispose in puncture-resistant containers.

10.10 Laboratory Documentation

Every laboratory should maintain proper records. Essential registers include:

Culture Register

Mother Culture Register

Substrate Preparation Register

Sterilization Log

Temperature Register

Humidity Register

Contamination Register

Harvest Register

Sales Register

Equipment Maintenance Register

Chemical Inventory Register

These records improve quality control and help identify problems.

10.11 Essential Equipment

The following table lists the basic equipment required for a small Cordyceps laboratory.

Equipment	Purpose	Essential for Beginners	Recommend for Commercial Labs
Pressure Cooker	Sterilization	✔	Backup
Autoclave	Sterilization	Optional	✔
Still Air Box	Sterile transfers	✔	Backup
Laminar Air Flow Cabinet	Sterile inoculation	Optional	✔
Alcohol Spray Bottle	Surface disinfection	✔	✔
Spirit Lamp/ Bunsen Burner	Flame sterilization	✔	✔
Electronic Balance	Accurate weighing	✔	✔
pH Meter	Media preparation	Recommended	✔
Hygrometer	Humidity monitoring	✔	✔

Equipment	Purpose	Essential for Beginners	Recomm for Comme Labs
Digital Thermometer	Temperature monitoring	✔	✔
LED Lighting	Fruiting induction	✔	✔
Refrigerator	Culture storage	✔	✔
Deep Freezer	Long-term preservation	Optional	Recomm

Estimated Laboratory Setup Cost (India)

Laboratory Type	Approximate Cost (₹)
Home Hobby Laboratory	25,000–50,000
Small Commercial Unit	1–3 lakh
Professional Laboratory	5–10 lakh
Research Laboratory	10 lakh and above

Note: These are indicative estimates and will vary with equipment quality, location, and scale.

Practical Tips

✔ Design the laboratory before purchasing equipment.

✔ Buy quality equipment rather than the cheapest available.

✔ Maintain separate footwear for the laboratory.

✔ Label every shelf and storage container.

✔ Keep a daily laboratory diary.

✔ Restrict visitor access to sterile areas.

✔ Inspect cultures at the same time each day.

Common Mistakes

⟐ Mixing dirty and clean work areas.

⟐ Opening culture bottles unnecessarily.

⟐ Ignoring small contamination spots.

⟐ Wearing outdoor shoes inside the laboratory.

⟐ Allowing insects to enter.

⟐ Failing to keep proper records.

⟐ Overcrowding the laboratory with unnecessary equipment.

Chapter Summary

In this chapter, you learned:

- How to choose an appropriate location for a Cordyceps laboratory.
- The importance of laboratory zoning and one-way workflow.
- The essential furniture and equipment required.
- Personal hygiene and sterile work practices.
- Waste management procedures.
- Record-keeping systems.
- Estimated setup costs for different scales of operation.

A well-planned laboratory greatly improves cultivation success, reduces contamination, and increases production efficiency.

Key Takeaways

✓ Cleanliness is more important than expensive equipment.

✓ Proper laboratory design minimizes contamination.

✓ Sterile techniques must become daily habits.

✓ Accurate documentation is essential for quality control.

✓ A disciplined grower consistently produces better results than an undisciplined one.

Chapter 11: Laboratory Equipment Selection, Installation, Operation, Calibration, and Maintenance

"Good equipment does not guarantee success, but poorly maintained equipment almost guarantees failure."

Learning Objectives

After completing this chapter, you will be able to:

- Select appropriate laboratory equipment according to your budget.
- Understand the purpose of each instrument.
- Install and operate equipment safely.
- Maintain equipment for long service life.
- Troubleshoot common operational problems.
- Estimate equipment costs for small and commercial laboratories.

11.1 Introduction

The quality of a Cordyceps laboratory depends not only on its cleanliness but also on the proper selection and maintenance of equipment. Every instrument in the laboratory has a specific purpose, and understanding its operation is essential for producing contamination-free cultures and high-quality fruiting bodies.

Beginners often purchase expensive equipment without understanding its function, while others attempt to cultivate Cordyceps with inadequate or poorly maintained tools. Both approaches lead to unnecessary failures.

This chapter explains the essential equipment required for successful cultivation, from simple hand tools to advanced laboratory instruments.

11.2 Pressure Cooker

Purpose

The pressure cooker is the most economical sterilization device for beginners. It is used for sterilizing:

- Culture media
- Brown rice substrate
- Glass bottles
- Culture jars
- Small laboratory tools

A pressure cooker can effectively destroy bacteria, mould spores, and other contaminants when used correctly.

Recommended Capacity

- 10–12 litres: Hobby cultivation
- 18–22 litres: Small production
- 30 litres or larger: Semi-commercial work

Operating Pressure

Maintain approximately **15 psi (103 kPa)**, which corresponds to about **121°C**, for effective sterilization.

Sterilization Time (Typical)

Material	Time
Liquid Media	20–25 minutes
Agar Media	20 minutes
Rice Bottles	90–120 minutes
Silkworm-Based Media	90–120 minutes
Laboratory Instruments	30 minutes

Common Mistakes

- Overfilling with water.
- Opening before pressure is fully released.
- Blocking the steam vent.
- Using damaged rubber gaskets.
- Ignoring the safety valve.

Maintenance

- Clean after every use.

- Replace worn gaskets.
- Check the pressure gauge regularly.
- Inspect safety valves monthly.
- Lubricate locking mechanisms if recommended by the manufacturer.

11.3 Autoclave

What is an Autoclave?

An autoclave is a specialized steam sterilizer designed for laboratory and medical applications.

Unlike domestic pressure cookers, autoclaves provide:

- Better temperature control
- Uniform sterilization
- Higher safety standards
- Greater capacity
- Automatic operation (in many models)

Commercial Cordyceps laboratories should use an autoclave whenever possible.

Types of Autoclaves

Vertical Autoclave

Suitable for:

- Small laboratories
- Educational institutions
- Research facilities

Horizontal Autoclave

Suitable for:

- Commercial mushroom farms
- Large research laboratories
- Industrial production

Typical Sterilization Cycle

1. Loading
2. Air removal
3. Steam generation
4. Sterilization
5. Pressure release
6. Drying
7. Cooling

Advantages

- ✔ Reliable sterilization
  - ✔ Large capacity
  - ✔ Automated control
  - ✔ Reduced contamination
  - ✔ Better reproducibility

11.4 Laminar Air Flow Cabinet

Importance

The Laminar Air Flow (LAF) cabinet is considered the heart of every modern fungal laboratory.

It provides a continuous stream of HEPA-filtered air that prevents airborne contaminants from reaching sterile cultures.

Without a clean inoculation environment, contamination rates increase dramatically.

Main Components

- HEPA filter
- Pre-filter
- Blower
- Fluorescent or LED lighting
- UV sterilization lamp

- Stainless steel work surface
- Transparent front shield

Operating Procedure

1. Clean the work surface with 70% alcohol.
2. Switch on the blower.
3. Allow the cabinet to run for 15–20 minutes.
4. Sterilize tools.
5. Perform inoculation with slow, deliberate movements.
6. Minimize talking and unnecessary hand movements.

Common Errors

- Blocking airflow with large objects.
- Working outside the sterile zone.
- Rapid hand movements.
- Opening culture bottles too long.
- Failing to replace filters when required.

11.5 Still Air Box (SAB)

For beginners, a Still Air Box offers an affordable alternative to a laminar flow cabinet.

It is simply a transparent plastic container with arm holes that reduces air movement during inoculation.

Although it does not filter the air, it minimizes airborne contamination when used correctly.

Advantages

- Very low cost
- Easy to construct
- Portable
- No electricity required

Limitations

- Slower operation
- Lower sterility than LAF cabinets
- Requires excellent technique

11.6 Electronic Balance

Accurate measurement is essential because even small errors in media composition can affect fungal growth.

Recommended Accuracy

- 0.01 g for routine laboratory work
- 0.001 g for research laboratories

Care

- Place on a vibration-free table.
- Protect from moisture.
- Calibrate regularly.
- Never overload the weighing pan.

11.7 pH Meter

The acidity (pH) of the culture medium influences nutrient availability and fungal growth. Most Cordyceps media perform best within a slightly acidic range, typically around **pH 5.5–6.5**, although the optimum depends on the formulation and strain.

Calibration

Calibrate the meter using standard buffer solutions (commonly pH 4.0, 7.0, and 10.0) according to the manufacturer's instructions.

11.8 Digital Thermometer

Temperature is one of the most critical factors affecting mycelial growth and fruiting. A reliable digital thermometer should be installed in:

- Incubation room
- Fruiting room
- Drying room
- Refrigerator

Record temperatures at least twice daily.

11.9 Digital Hygrometer

Humidity influences:

- Mycelial growth
- Primordia formation
- Fruiting body development

- Product quality

Install hygrometers in all cultivation rooms and verify their accuracy periodically.

11.10 Refrigerator and Freezer

Proper culture storage extends the life of valuable strains.

Refrigerator (2–8°C)

Used for:

- Mother cultures
- Media
- Reagents
- Short-term storage

Freezer

Used for:

- Long-term preservation of selected materials
- Certain laboratory reagents

Avoid repeated freeze-thaw cycles unless specifically appropriate for the material being stored.

11.11 Microscope

Although beginners may not use a microscope every day, it becomes invaluable for:

- Identifying contamination
- Examining spores
- Checking culture purity
- Educational demonstrations

A compound microscope with magnifications up to **400×** is suitable for most routine work.

11.12 LED Lighting System

Unlike green plants, Cordyceps does not require light for photosynthesis. However, light acts as an environmental signal that helps regulate normal fruiting body development and pigmentation.

For most cultivation systems:

- Cool white LED lighting is commonly used.
- Moderate intensity is generally sufficient.
- Uniform illumination is preferable to bright spotlights.
- Excessive heat from lighting should be avoided.

Equipment Maintenance Schedule

Equipment	**Daily**	**Weekly**	**Monthly**	**Yearly**
Pressure Cooker	✔	✔	✔	Replace gasket if needed
Autoclave	✔	✔	✔	Professic servicing
Laminar Flow Cabinet	✔	✔	Inspect filters	HEPA filt evaluatic
Electronic Balance	✔	Calibration check	✔	Professic calibratic
pH Meter	Clean	Buffer calibration	Electrode inspection	Replace electrod required
Thermometer	✔	Compare readings	✔	Replace battery
Hygrometer	✔	Compare readings	✔	Replace battery

Estimated Equipment Cost (India)

Equipment	Approximate Price (₹)
Pressure Cooker	4,000–8,000
Vertical Autoclave	50,000–1,50,000
Laminar Air Flow Cabinet	35,000–90,000
Still Air Box	1,000–3,000
Electronic Balance	2,000–15,000
pH Meter	2,500–10,000
Hygrometer	500–2,000
Digital Thermometer	300–1,500
Refrigerator	18,000–35,000
Compound Microscope	12,000–60,000

Prices are indicative and may vary depending on the manufacturer, specifications, and market conditions.

Practical Tips

✓ Buy laboratory-grade equipment whenever possible.

✓ Keep operating manuals in a dedicated file.

✓ Label every instrument with its calibration date.

✓ Train all staff before allowing them to operate equipment.

✓ Maintain a preventive maintenance schedule rather than waiting for equipment to fail.

✓ Always keep spare items such as gaskets, fuses, thermometer batteries, alcohol, and filter paper.

Common Mistakes to Avoid

⟐ Purchasing equipment without considering future expansion.

⟐ Using domestic appliances beyond their intended capacity.

⟐ Ignoring calibration and maintenance.

⟐ Operating sterilization equipment without adequate safety precautions.

⟐ Storing cultures in refrigerators containing food items.

◈ Failing to maintain equipment logs.

Chapter Summary

In this chapter, you learned:

- The purpose and operation of the essential equipment used in a *Cordyceps militaris* laboratory.
- How to select equipment appropriate to your production scale.
- Safe operating practices.
- Routine maintenance and calibration procedures.
- Indicative costs for establishing a laboratory in India.

Understanding your equipment is as important as understanding the fungus itself. Well-maintained instruments improve consistency, reduce contamination, and contribute to high-quality production.

Chapter 12: Raw Materials and Culture Media Selection, Preparation, and Quality Control

"The quality of a Cordyceps crop is determined long before inoculation. It begins with selecting the right raw materials and preparing a nutritionally balanced, contamination-free culture medium."

Learning Objectives

After studying this chapter, you will be able to:

- Identify the essential raw materials required for *Cordyceps militaris* cultivation.
- Understand the nutritional role of each ingredient.
- Select high-quality materials from reliable sources.
- Prepare different substrate formulations for home, research, and commercial cultivation.
- Maintain consistency through proper quality control.

12.1 Introduction

Successful cultivation of *Cordyceps militaris* depends on two fundamental requirements:

1. A healthy and vigorous fungal culture.
2. A well-balanced nutrient medium.

Even the best fungal strain cannot produce high-quality fruiting bodies if the substrate lacks essential nutrients or becomes contaminated. Likewise, using premium ingredients without proper sterilization or mixing techniques can result in poor colonization and crop failure. This chapter explains the function of each major ingredient, how to prepare cultivation media, and how to maintain quality from batch to batch.

12.2 Characteristics of an Ideal Culture Medium

A good substrate should:

- Supply balanced nutrients.
- Retain adequate moisture.
- Allow proper gas exchange.

- Support vigorous mycelial growth.
- Promote uniform fruiting.
- Resist contamination after sterilization.
- Be economical and readily available.

No single formulation is suitable for every strain. Small adjustments are often needed based on local conditions and cultivation goals.

12.3 Water – The Most Important Ingredient

Although often overlooked, water is the largest component of the culture medium.

Characteristics of Good Water

- Clean
- Colourless
- Odourless
- Free from excessive chlorine
- Low microbial contamination
- Near-neutral pH (approximately 6.5–7.5)

Recommended Sources

✔ RO (Reverse Osmosis) water

✔ Distilled water (for laboratory work)

✔ Clean drinking water

Avoid heavily mineralized or contaminated water, as it may inhibit fungal growth.

12.4 Brown Rice

Brown rice is the most widely used substrate for *Cordyceps militaris* because it provides carbohydrates, vitamins, minerals, and a suitable physical structure.

Advantages

- Easily available
- Affordable
- Consistent quality
- Excellent texture
- Good moisture retention
- Supports high yields

Selection Criteria

Choose rice that is:

- Fresh

- Uniform in size
- Free from insects
- Free from mould
- Free from chemical odours

Do not use old, broken, or mouldy grains.

12.5 White Rice vs. Brown Rice

Feature	**Brown Rice**	**White Rice**
Nutritional Value	Higher	Lower
Fibre Content	Higher	Lower
Vitamin Content	Higher	Lower
Mineral Content	Higher	Lower
Preferred for Cordyceps	✔	Occasionally

Brown rice generally provides a more balanced nutritional profile, making it the preferred choice for most cultivation systems.

12.6 Silkworm Pupae Powder

Silkworm pupae powder is a valuable protein-rich supplement used in many commercial formulations.

It contributes:

- Proteins
- Amino acids
- Lipids
- Vitamins
- Trace minerals

These nutrients support vigorous mycelial growth and fruiting body development.

Quality Requirements

- Fresh
- Finely ground
- Low moisture
- Free from unpleasant odours
- Properly stored

12.7 Yeast Extract

Yeast extract supplies:

- B vitamins
- Amino acids
- Peptides
- Growth factors

Although relatively expensive, it significantly improves nutrient availability in many laboratory media.

12.8 Peptone

Peptone provides readily available nitrogen.

Nitrogen is essential for:

- Protein synthesis
- Enzyme production
- Mycelial development

Only laboratory-grade peptone should be used for research applications.

12.9 Glucose

Glucose is an immediately available carbon source.

Benefits include:

- Rapid initial growth
- Improved energy supply
- Faster mycelial colonization

Excessive glucose, however, may encourage contamination if sterilization is inadequate.

12.10 Sucrose

Sucrose can also serve as an energy source.

Some growers combine glucose and sucrose to provide a balanced carbohydrate profile.

12.11 Mineral Supplements

Although required only in small quantities, minerals play important roles in fungal metabolism.

Commonly used minerals include:

- Magnesium sulfate

- Potassium phosphate
- Calcium carbonate
- Zinc sulfate
- Iron salts

Avoid excessive mineral supplementation, as imbalance may inhibit growth.

12.12 Vitamin Supplements

Cordyceps usually obtains sufficient vitamins from rice, yeast extract, and other ingredients.

In research laboratories, defined vitamin mixtures may be added to specialized media for experimental purposes.

12.13 Agar

Agar is used primarily for preparing solid media in Petri dishes.

It is extracted from marine algae and serves as a solidifying agent.

Agar itself provides little nutrition but creates a firm surface for fungal growth.

12.14 Culture Bottles

Glass bottles remain the preferred container for fruiting-body cultivation.

Advantages

- Reusable
- Heat resistant
- Easy to sterilize
- Environmentally friendly

Recommended bottle volume:

300–500 ml for routine production.

12.15 Polypropylene Containers

Many commercial farms use polypropylene bottles because they are:

- Lightweight
- Durable
- Reusable
- Resistant to high sterilization temperatures

Always ensure they are autoclavable.

12.16 Cotton Plugs and Filter Caps

Proper closures allow gas exchange while preventing contamination.

Common options include:

- Cotton plugs
- Synthetic filter plugs
- Filter caps with microporous membranes

Filter caps provide better consistency in commercial cultivation.

12.17 Basic Culture Media Recipes

Formula A – Beginner's Rice Medium

- Brown rice – 50 g
- Water – 60 ml

Simple, economical, and ideal for learning basic cultivation techniques.

Formula B – Nutrient-Enriched Medium

- Brown rice – 50 g
- Silkworm pupae powder – 5 g

- Yeast extract – 2 g
- Water – 60 ml

Provides improved nutrition for vigorous growth.

Formula C – Commercial Production Medium

- Brown rice – 45 g
- Silkworm pupae powder – 8 g
- Yeast extract – 2 g
- Mineral supplement – As recommended
- Water – Adjust to the desired moisture content

Commercial producers often optimize this formulation through small-scale trials before large-scale production.

12.18 Moisture Content

Moisture is one of the most critical factors.

Too Little Water

- Slow colonization
- Poor nutrient utilization

- Reduced yield

Too Much Water

- Sticky substrate
- Bacterial contamination
- Poor aeration
- Delayed fruiting

Proper moisture balance is essential for healthy growth.

12.19 Mixing Procedure

1. Weigh all ingredients accurately.
2. Mix dry ingredients thoroughly.
3. Add water gradually while stirring.
4. Ensure uniform moisture distribution.
5. Fill bottles without compressing the substrate.
6. Wipe bottle rims clean before capping.

Uniform preparation leads to more consistent results.

12.20 Quality Control During Media Preparation

Every batch should be checked for:

- Correct ingredient weights
- Moisture consistency
- Uniform mixing
- Bottle cleanliness
- Proper labeling
- Preparation date
- Batch number

Maintaining detailed records allows growers to identify and correct problems in future batches.

Common Problems and Solutions

Problem	Possible Cause	Solution
Slow mycelial growth	Poor nutrition or incorrect moisture	Review the formulation and moisture level
Wet substrate	Excess water	Reduce water content and improve mixing
Dry substrate	Insufficient water	Add water gradually in future batches
Uneven growth	Poor mixing	Mix ingredients more thoroughly
Frequent contamination	Contaminated ingredients or poor sterilization	Improve raw material quality and sterilization procedures

Practical Tips

✔ Purchase raw materials from trusted suppliers.

✔ Store ingredients in airtight containers.

✔ Label every batch with the preparation date.

✔ Use the oldest stock first (First-In, First-Out method).

✔ Conduct small trial batches before changing formulations.

✔ Keep a written record of every substrate recipe and its performance.

Chapter Summary

In this chapter, you learned:

- The role of each major raw material used in *Cordyceps militaris* cultivation.
- How to select high-quality ingredients.
- The nutritional importance of brown rice, silkworm pupae powder, yeast extract, and mineral supplements.
- Practical substrate formulations for beginners and commercial growers.
- The importance of moisture control, uniform mixing, and batch documentation.

A well-prepared culture medium provides the nutritional foundation for healthy mycelial growth and successful fruiting.

Key Takeaways

✓ High-quality ingredients produce healthier cultures.

✓ Consistency in media preparation leads to consistent production.

✓ Moisture management is one of the most important factors in cultivation success.

✓ Good record-keeping helps improve production efficiency over time.

Chapter 13: Pure Culture Techniques

Isolation, Tissue Culture, Agar Media Preparation, and Mother Culture Maintenance

"The future of every Cordyceps crop is determined by the purity of its culture. One contaminated culture can destroy months of hard work, while one healthy mother culture can produce thousands of successful bottles."

Learning Objectives

After completing this chapter, you will be able to:

- Understand the concept of pure fungal culture.
- Learn different methods of isolating *Cordyceps militaris*.
- Prepare commonly used agar media.
- Perform tissue culture and culture transfer techniques.
- Maintain healthy mother cultures.
- Identify and eliminate contamination.
- Preserve cultures for long-term use.

13.1 Introduction

Pure culture technology forms the foundation of scientific mushroom cultivation. Every commercial Cordyceps farm, biotechnology laboratory, and research institution depends upon genetically pure and contamination-free fungal cultures.

Without a healthy mother culture:

- Mycelial growth becomes weak.
- Contamination increases.
- Fruiting becomes irregular.
- Yield decreases.
- Cordycepin production may decline.
- Commercial losses become inevitable.

Therefore, learning aseptic culture techniques is one of the most valuable skills for every Cordyceps cultivator.

13.2 What Is a Pure Culture?

A **pure culture** contains only one fungal strain growing without contamination from bacteria, yeasts, moulds, or other fungi.

Characteristics of a good pure culture include:

- Uniform white mycelium
- Vigorous growth
- No coloured patches
- No foul smell
- No bacterial slime
- Regular colony margins
- Stable growth rate

A healthy pure culture is the starting point for all subsequent cultivation.

13.3 Types of Cultures Used in Cordyceps Cultivation

Commercial laboratories generally maintain several levels of cultures.

Master Culture

The original authenticated strain.

Characteristics:

- Highest genetic purity
- Rarely used directly
- Stored under secure conditions
- Replaced only when necessary

Mother Culture

Prepared from the master culture.

Used for:

- Routine laboratory work
- Preparation of spawn
- Production cultures

Working Culture

Prepared from the mother culture.

Used daily for:

- Bottle inoculation
- Experimental work
- Training

- Demonstrations

Maintaining this hierarchy reduces the risk of losing valuable strains.

13.4 Sources of Pure Cultures

Reliable cultures should be obtained from:

- National culture collections
- Universities
- Government research institutes
- Accredited biotechnology laboratories
- Reputable commercial suppliers

Avoid purchasing unidentified cultures from unverified online sources.

13.5 Laboratory Sterility Before Culture Work

Before opening any culture:

Clean the Workspace

Disinfect with 70% ethanol or isopropyl alcohol.

Sterilize Instruments

Flame sterilize:

- Forceps
- Scalpels
- Inoculation needles

Allow tools to cool before touching fungal tissue.

Personal Protection

Wear:

- Laboratory coat
- Face mask
- Gloves
- Hair cover (if required)

Air Sterility

Operate the Laminar Air Flow cabinet for at least **15–20 minutes** before beginning work, or prepare a properly disinfected Still Air Box.

13.6 Agar Media Used for Cordyceps

Solid media provide a surface for fungal growth.

The most commonly used agar media include:

- Potato Dextrose Agar (PDA)
- Sabouraud Dextrose Agar with Yeast Extract (SDAY)
- Malt Extract Agar (MEA)
- Oatmeal Agar (OA)

Each medium has specific advantages.

13.7 Potato Dextrose Agar (PDA)

PDA is widely used because it is simple, economical, and supports vigorous fungal growth.

Typical Ingredients (Per Litre)

- Potato infusion
- Dextrose
- Agar
- Distilled water

After preparation, sterilize before pouring into sterile Petri dishes.

PDA is particularly useful for:

- Culture maintenance
- Colony observation
- Routine transfers

13.8 SDAY Medium

Sabouraud Dextrose Agar with Yeast Extract is one of the preferred media for **Cordyceps militaris**.

Advantages

- Rich nutrient composition
- Excellent mycelial growth
- High colony uniformity
- Suitable for laboratory research

Many commercial laboratories routinely use SDAY for mother culture maintenance.

13.9 Agar Preparation Procedure

Step 1

Prepare the medium according to the selected formulation.

Step 2

Adjust pH if necessary.

Step 3

Heat until agar dissolves completely.

Step 4

Dispense into suitable bottles or flasks.

Step 5

Sterilize at approximately **121°C (15 psi)** for about **15–20 minutes**.

Step 6

Allow the medium to cool to around **45–50°C** before pouring into sterile Petri dishes inside a sterile workspace.

Step 7

Leave dishes undisturbed until the agar solidifies.

13.10 Tissue Culture Technique

Tissue culture is the preferred method for obtaining cultures from a healthy fruiting body.

Selection of Fruiting Body

Choose specimens that are:

- Healthy
- Fully developed
- Disease-free
- Fresh
- Free from contamination

Avoid damaged or over-mature fruiting bodies.

Surface Sterilization

Before excising tissue:

- Remove external debris.

- Briefly disinfect the surface using an appropriate sterilization protocol.
- Work aseptically to minimize contamination.

Tissue Excision

Using a sterile scalpel:

- Remove a small piece of internal tissue from the centre of the fruiting body.
- Avoid the outer surface, which is more likely to carry contaminants.

Transfer

Place the tissue gently onto the prepared agar medium.

Seal the Petri dish and incubate under suitable conditions.

13.11 Culture Transfer (Subculturing)

As the colony grows, transfer healthy mycelium to fresh agar.

Procedure

1. Sterilize the inoculation tool.
2. Open the culture plate only briefly.
3. Cut a small block from the actively growing edge of the colony.
4. Transfer it to a fresh plate.
5. Close and seal immediately.

Always use the leading edge of the colony because it contains the youngest, most vigorous mycelium.

13.12 Incubation

Typical incubation conditions include:

Parameter	Recommended Range
Temperature	20–24°C
Light	Darkness or very low light
Humidity	Normal laboratory conditions
Disturbance	Minimum

Inspect cultures daily without opening the plates.

13.13 Recognizing Healthy Mycelium

Healthy *Cordyceps militaris* mycelium generally appears:

- Bright white
- Cottony to dense
- Uniform
- Radially expanding
- Free from discoloration

Growth characteristics may vary slightly between strains.

13.14 Recognizing Contamination

Bacterial Contamination

Signs include:

- Wet appearance
- Sticky colonies
- Yellowish slime
- Sour odour

Green Mould (*Trichoderma*)

Characteristics:

- Rapid growth
- Initially white
- Turns green after sporulation
- Highly destructive

Black Mould

Usually appears as dark or black colonies.

Should be discarded immediately.

Yeast Contamination

May appear as:

- Cream-coloured colonies
- Smooth surface
- Shiny texture

13.15 What Should Be Done With Contaminated Cultures?

Never attempt to "save" contaminated cultures.

Recommended procedure:

- Seal the container.
- Sterilize it before disposal.

- Clean the work area.
- Review possible sources of contamination.
- Record the incident in the contamination log.

13.16 Mother Culture Maintenance

A mother culture should receive careful attention.

Maintain:

- Proper labeling
- Date of preparation
- Batch number
- Strain identification
- Transfer history

Avoid excessive subculturing, as repeated transfers may lead to reduced vigor or genetic changes over time.

13.17 Long-Term Preservation

Several methods are used for preserving valuable strains.

Refrigeration (2–8°C)

Suitable for short-term storage.

Mineral Oil Preservation

Cultures are covered with sterile mineral oil to slow growth.

Useful for medium-term preservation.

Cryopreservation

Storage at ultra-low temperatures using specialized equipment.

Provides excellent long-term genetic stability.

Typically used in research institutions.

Lyophilization (Freeze-Drying)

Used by culture collections for long-term preservation of selected microorganisms.

Not routinely used in most Cordyceps production laboratories.

13.18 Laboratory Records

Maintain detailed records including:

- Culture number

- Source
- Date received
- Date transferred
- Medium used
- Incubation conditions
- Observations
- Contamination incidents

Accurate documentation improves traceability and quality assurance.

Troubleshooting Guide

Problem	Possible Cause	Suggested Solution
No growth	Non-viable culture or incorrect incubation conditions	Verify culture viability and environmental conditions
Slow growth	Old culture or unsuitable medium	Prepare fresh medium and use a vigorous culture
Repeated contamination	Poor aseptic technique	Review sterile procedures and disinfect the workspace thoroughly
Irregular colonies	Mixed culture or strain degeneration	Re-isolate from a healthy culture and minimize repeated subculturing
Agar drying out	Poor storage or prolonged	Seal plates properly and

Problem	Possible Cause	Suggested Solution
	incubation	incubate only as long as necessary

Practical Tips

✓ Work slowly and deliberately during culture transfers.

✓ Label every Petri dish before inoculation.

✓ Never rely on memory—record every transfer.

✓ Keep master cultures separate from working cultures.

✓ Use fresh agar media for important transfers.

✓ Inspect cultures daily without opening the plates.

✓ Dispose of contaminated cultures promptly and safely.

Chapter Summary

In this chapter, you learned:

- The importance of pure cultures in

Cordyceps militaris cultivation.

- The hierarchy of master, mother, and working cultures.
- Preparation and use of common agar media.
- Tissue culture and subculturing techniques.
- Recognition and management of contamination.
- Methods for preserving valuable fungal strains.
- The importance of meticulous laboratory records.

A contamination-free culture is the cornerstone of successful Cordyceps cultivation. Investing time in mastering aseptic techniques will improve productivity, reduce losses, and ensure consistent crop quality.

Key Takeaways

✓ Pure cultures are the foundation of every successful cultivation programme.

✓ Aseptic technique is a practical skill that improves with consistent practice.

✓ Early detection of contamination prevents major production losses.

✓ Proper culture preservation protects valuable strains for future use.

Chapter 14 : Spawn Production and Liquid Culture Technology for Cordyceps militaris

"Healthy spawn is the seed of successful cultivation. Every vigorous fruiting body begins with vigorous mycelium."

Learning Objectives

After completing this chapter, you will be able to:

- Understand the importance of spawn in Cordyceps cultivation.
- Differentiate between solid spawn and liquid culture.
- Prepare high-quality grain spawn.
- Produce contamination-free liquid cultures.
- Evaluate spawn quality before inoculation.
- Scale up inoculum production for commercial cultivation.
- Troubleshoot common spawn-related problems.

14.1 Introduction

Unlike green plants, fungi are not grown from seeds. Instead, they are propagated using **mycelium**, the vegetative part of the fungus. When mycelium is transferred onto a nutritious substrate under sterile conditions, it colonizes the medium and eventually produces fruiting bodies. In mushroom cultivation, this actively growing mycelium on a carrier material is called **spawn**. Spawn serves as the "planting material" for fungal cultivation. For *Cordyceps militaris*, the quality of spawn directly affects:

- Colonization speed
- Contamination rate
- Fruiting body formation
- Yield
- Uniformity
- Bioactive compound production

Commercial laboratories often describe spawn quality as the single most important determinant of cultivation success.

14.2 What is Spawn?

Spawn is a sterile substrate colonized by healthy *Cordyceps militaris* mycelium.

It acts as the inoculum used to introduce the fungus into the final production substrate.

A good spawn should:

- Be genetically pure.
- Be free from contaminants.
- Show vigorous mycelial growth.
- Colonize quickly.
- Have a pleasant mushroom aroma.
- Be correctly labeled with strain and preparation date.

14.3 Types of Spawn

Different cultivation systems use different types of spawn.

Grain Spawn

Prepared on cereal grains.

Most common for laboratory and commercial use.

Liquid Spawn

Prepared in nutrient broth.

Increasingly used for large-scale production because it allows rapid inoculation.

Sawdust Spawn

Commonly used for wood-decaying mushrooms but less common for *Cordyceps militaris.*

Agar Spawn

Used mainly for laboratory transfers rather than commercial production.

14.4 Grain Selection

Several grains have been successfully used. Examples include:

- Wheat
- Sorghum (Milo)
- Pearl millet (Bajra)
- Barley
- Rye
- Brown rice

The choice depends on:

- Local availability
- Cost
- Ease of sterilization
- Mycelial growth performance

14.5 Characteristics of Good Grain

Select grains that are:

✔ Fresh

✔ Mature

✔ Uniform in size

✔ Free from insects

✔ Free from mould

✔ Properly dried

✔ Free from pesticide residues

Avoid cracked, broken, or mouldy grains.

14.6 Grain Preparation

Step 1: Cleaning

Remove:

- Dust

- Stones
- Broken grains
- Plant debris

Wash thoroughly with clean water.

Step 2: Soaking

- Soak grains for **8–12 hours** (or according to the grain type) to allow adequate water absorption.

Step 3: Boiling

- Simmer the grains until they are hydrated but still intact. Properly prepared grains should be firm and not burst easily.

Step 4: Draining

- Drain excess water thoroughly.
- The grain surface should appear dry while the interior retains moisture.
- Excess surface moisture increases the risk of bacterial contamination.

14.7 Moisture Content

Correct moisture is essential.

Too Dry

- Slow colonization
- Weak mycelial growth

Too Wet

- Bacterial contamination
- Grain clumping
- Poor aeration

Aim for evenly hydrated grains without free water.

14.8 Supplementation

Some laboratories add small amounts of supplements such as gypsum or calcium carbonate to improve grain separation and help stabilize pH. Supplementation should be standardized and validated within each production system.

14.9 Filling Spawn Bottles

Fill bottles approximately **two-thirds full**, leaving sufficient air space for gas exchange and shaking.

Label each bottle with:

- Batch number
- Grain type
- Date of preparation
- Strain identification

14.10 Sterilization

Sterilize the prepared grain using an autoclave or pressure cooker.

Typical conditions are:

- Approximately **121°C (15 psi)**
- **90–120 minutes**, depending on bottle size and load

Allow the bottles to cool completely before inoculation.

14.11 Inoculation

Transfer a small amount of healthy mycelium from a fresh agar culture into the sterilized grain under aseptic conditions.

Use either:

- Laminar Air Flow Cabinet
- Still Air Box

Minimize the time containers remain open.

14.12 Incubation

Typical incubation conditions:

Parameter	Recommended Range
Temperature	20–24°C
Light	Dark or low light
Humidity	Ambient laboratory conditions
Disturbance	Minimal

Inspect bottles daily without opening them.

14.13 Shaking Spawn

When the grain is partially colonized (often around 25–40%), gently shake the bottle to redistribute the mycelium.

Benefits include:

- Faster colonization
- More uniform growth
- Easier detection of contamination

Avoid excessive shaking, which can damage delicate mycelial networks.

14.14 Characteristics of Good Spawn

Healthy spawn should have:

- Bright white mycelium
- Uniform colonization
- Pleasant mushroom aroma
- No wet patches
- No green, black, pink, or yellow colonies
- No bacterial slime

14.15 What is Liquid Culture?

Liquid culture is a sterile nutrient solution containing actively growing fungal mycelium.

Unlike grain spawn, the fungus grows suspended within the liquid medium.

Liquid culture allows rapid multiplication of inoculum.

14.16 Advantages of Liquid Culture

✔ Faster inoculation

✔ Uniform distribution

✔ Easier scale-up

✔ Reduced preparation time

✔ Lower labour requirement

✔ Suitable for commercial automation

14.17 Liquid Culture Medium

A typical liquid medium contains:

- Carbon source
- Nitrogen source
- Mineral salts
- Vitamins
- Sterile water

The exact formulation varies among laboratories and should be optimized for the selected strain.

14.18 Preparing Liquid Culture

1. Prepare the nutrient solution.
2. Adjust pH if required.
3. Dispense into sterile flasks.
4. Sterilize.
5. Cool completely.
6. Inoculate aseptically.
7. Incubate under suitable conditions.

14.19 Agitation

Many laboratories use magnetic stirrers or orbital shakers to maintain gentle movement during incubation.

Benefits include:

- Improved oxygen transfer
- Even nutrient distribution
- Uniform mycelial growth
- Reduced clumping

14.20 Evaluating Liquid Culture

A healthy liquid culture typically appears as suspended white mycelial strands or small aggregates in a clear nutrient solution.

Discard cultures showing:

- Cloudiness unrelated to mycelial growth
- Unpleasant odours
- Coloured contamination
- Gas production
- Surface films suggestive of bacterial or yeast contamination

14.21 Scaling Up Production

Commercial facilities often follow a staged approach:

Master Culture → Mother Culture → Grain Spawn or Liquid Culture → Production Bottles

This system minimizes contamination risk and preserves valuable strains.

14.22 Quality Control Checklist

Before using spawn:

✓ Confirm strain identity.

✓ Verify complete colonization.

✓ Check for contamination.

✓ Confirm labeling accuracy.

✓ Review incubation records.

✓ Ensure the spawn is within the recommended storage period.

Troubleshooting Guide

Problem	Possible Cause	Solution
Slow colonization	Old culture or poor-quality grain	Use fresh cultures and high-quality grain
Wet spawn	Excess moisture	Improve grain draining before sterilization
Green mould	Contamination during inoculation	Review aseptic technique
Sour smell	Bacterial contamination	Discard the batch and investigate the source
Uneven growth	Inadequate shaking or inconsistent hydration	Standardize grain preparation and shaking

Practical Tips

✔ Prepare small trial batches before large-scale production.

✔ Never mix old and new spawn.

✔ Maintain separate storage for master cultures and production spawn.

✔ Record every batch in a spawn register.

✔ Rotate stock using the "First In, First Out (FIFO)" principle.

✔ Train laboratory personnel in aseptic handling.

Chapter Summary

In this chapter, you learned:

- The role of spawn in *Cordyceps militaris* cultivation.
- How to prepare high-quality grain spawn.
- The principles of liquid culture production.
- Proper incubation, shaking, and quality evaluation.
- Procedures for scaling up inoculum production.
- Common problems and their solutions.

Healthy spawn is the foundation of every successful cultivation cycle. By maintaining

15.1 Introduction

Bottle cultivation has become the preferred production system for *Cordyceps militaris* throughout the world. Compared with cultivation in plastic bags or trays, bottle cultivation offers several important advantages:

- Better contamination control
- Uniform substrate volume
- Improved moisture retention
- Easier handling
- Higher-quality fruiting bodies
- Greater consistency between batches
- Reusable containers that reduce long-term costs

Whether the production scale is 100 bottles or 100,000 bottles, the same scientific principles apply.

15.2 Choosing the Right Bottle

An ideal cultivation bottle should possess the following characteristics:

- Heat resistant
- Easy to sterilize
- Transparent
- Durable
- Chemically inert
- Reusable
- Wide mouth for easy harvesting

Recommended Bottle Sizes

Bottle Size	Suitable For
250 ml	Research work
350 ml	Small-scale cultivation
500 ml	Commercial cultivation
750 ml	Experimental production

The **350–500 ml bottle** is considered ideal for most commercial growers.

15.3 Glass vs. Polypropylene Bottles

Feature	Glass Bottle	Polypropylene Bottle
Reusable	Excellent	Excellent
Heat Resistance	Excellent	Excellent (Autoclavable Grade)
Breakage	Possible	Rare
Cost	Moderate	Moderate
Weight	Heavy	Light
Commercial Use	Widely Used	Increasingly Popular

15.4 Bottle Cleaning

Never reuse bottles without proper cleaning.

Cleaning Procedure

1. Remove old substrate completely.
2. Wash using warm water and a mild laboratory detergent.
3. Scrub thoroughly.
4. Rinse several times with clean water.
5. Allow bottles to dry.

6. Inspect each bottle for cracks or damage.

Discard damaged bottles.

15.5 Preparing the Production Substrate

A typical production substrate consists of:

- Brown rice
- Nutritional supplements (if used)
- Clean water

All ingredients should be weighed accurately.

Example Production Formula (Per Bottle)

- Brown rice – 50 g
- Silkworm pupae powder – 5 g
- Yeast extract – 2 g
- Mineral supplement – As required
- Water – Approximately 60 ml (adjusted according to the formulation)

Each laboratory should standardize its own validated formulation.

15.6 Mixing Procedure

1. Weigh all ingredients.
2. Mix dry ingredients thoroughly.
3. Add water gradually.
4. Ensure even moisture distribution.
5. Avoid lumps.
6. Use immediately after preparation.

Never prepare more substrate than can be sterilized promptly.

15.7 Filling the Bottles

Fill bottles to approximately **one-third to one-half** of their total volume. Avoid overfilling, as adequate headspace is required for gas exchange and fruiting body development.

Do not compress the substrate tightly.

Level the surface gently.

Clean the bottle rim before capping.

15.8 Bottle Closures

Several closure systems are used:

Cotton Plug

Advantages:

- Low cost
- Easy to prepare

Disadvantages:

- Variable air exchange
- Requires careful handling

Filter Cap

Commercial laboratories generally prefer filter caps because they provide:

- Better sterility
- Controlled gas exchange
- Reduced contamination
- Improved consistency

Microporous Membrane Cap

Used in advanced laboratories.

Provides excellent filtration.

15.9 Labeling

Each bottle should be labeled with:

- Batch Number
- Date
- Medium Formula
- Strain Number
- Operator Initials

Proper labeling prevents confusion during production.

15.10 Sterilization

Sterilization destroys microorganisms before inoculation.

Standard Conditions

- Temperature: **121°C**
- Pressure: **15 psi**
- Duration: **90–120 minutes** (depending on bottle size and load)

Do not shorten sterilization time without validation.

15.11 Cooling

After sterilization:

- Allow bottles to cool naturally.
- Do not remove lids during cooling.
- Avoid rapid cooling with water or fans.

Bottles should reach room temperature before inoculation.

15.12 Inoculation Methods

Several inoculation methods are practiced.

Agar Block Inoculation

A small agar block from a healthy mother culture is transferred into each bottle.

Advantages:

- Reliable
- Simple
- Widely used

Grain Spawn Inoculation

Small quantities of colonized grain are introduced into production bottles.

Advantages:

- Rapid colonization
- Easy scale-up

Liquid Culture Inoculation

Sterile syringes or dispensing systems introduce liquid mycelium into the substrate.

Advantages:

- Fast
- Uniform
- Suitable for commercial automation

15.13 Aseptic Technique During Inoculation

During inoculation:

✔ Wear gloves.

✔ Wear a face mask.

✔ Sterilize tools before each transfer.

✔ Work close to the sterile airflow.

✔ Minimize exposure time.
✔ Keep movements slow and deliberate.
✔ Avoid talking unnecessarily.

15.14 After Inoculation

Immediately after inoculation:

- Close bottles securely.
- Inspect labels.
- Transfer bottles to the incubation room.
- Arrange bottles neatly on racks.

Avoid shaking production bottles unless your validated cultivation protocol specifically requires it.

15.15 Incubation Conditions

Typical incubation conditions:

Parameter	Recommended Range
Temperature	20–24°C
Relative Humidity	Ambient laboratory conditions
Light	Darkness or very low light
Air Movement	Minimal
Disturbance	Minimal

Inspect bottles visually every day.

Do not open bottles during incubation.

15.16 Colonization Stages

Healthy bottles generally pass through the following stages:

Stage 1

Small white mycelial patches appear.

Stage 2

Mycelium spreads across the substrate.

Stage 3

Most of the substrate becomes colonized.

Stage 4

Complete colonization.

Bottle becomes ready for fruiting induction.

15.17 Quality Control Checklist

Before transferring bottles for fruiting:

✓ Correct labeling

✓ Complete colonization

✓ No contamination

✓ Uniform colour

✓ Pleasant mushroom aroma

✓ Proper moisture

15.18 Production Workflow

Raw Materials

|

▼

Media Preparation

|

▼

Bottle Filling

|

▼

Sterilization

|

▼

Cooling

|

▼

Inoculation

|

▼

Incubation

|

▼

Complete Colonization

|

▼

Transfer to Fruiting Room

Standard Operating Procedure (SOP)

Before Starting

- Wash hands.
- Wear laboratory clothing.
- Disinfect work surfaces.
- Check equipment.
- Verify culture identity.

During Inoculation

- Maintain sterile conditions.
- Label bottles immediately.
- Minimize bottle opening time.
- Sterilize tools between transfers.

After Completion

- Clean the work area.
- Record batch details.
- Dispose of waste safely.
- Inspect bottles the following day.

Common Problems

Problem	Cause	Solution
Bacterial contamination	Inadequate sterilization or poor aseptic technique	Review sterilization cycle and inoculation procedures
Uneven colonization	Inconsistent substrate preparation	Standardize mixing and filling
Dry substrate	Insufficient water	Adjust moisture in future batches
Wet substrate	Excess water	Reduce water and improve drainage
Label loss	Poor adhesive	Use waterproof laboratory labels

Practical Tips

✔ Standardize every step of bottle preparation.

✔ Prepare written SOPs for staff.

✔ Work in small batches when learning.

✔ Use a contamination logbook.

✔ Calibrate sterilization equipment regularly.

✔ Photograph each production batch for documentation.

✔ Rotate bottles to maintain production scheduling.

Commercial Production Checklist

Before starting a production batch, verify:

- Raw materials available.
- Cultures healthy.
- Bottles clean.
- Sterilizer functioning correctly.
- Laminar airflow cabinet disinfected.
- Labels prepared.
- Production register updated.
- Emergency supplies available.

Chapter Summary

In this chapter, you learned:

- How to select suitable cultivation bottles.
- Proper substrate preparation and bottle filling techniques.

- Standard sterilization procedures.
- Safe and effective inoculation methods.
- Incubation practices following inoculation.
- Standard Operating Procedures for commercial production.
- Quality-control measures that improve consistency and reduce contamination.

By carefully following these procedures, cultivators can establish a reliable production system that produces healthy, contamination-free bottles ready for fruiting.

Key Takeaways

✓ Uniform bottle preparation leads to uniform crop performance.

✓ Sterilization is essential for contamination control.

✓ Accurate labeling and documentation improve traceability.

✓ Consistent SOPs are the hallmark of successful commercial cultivation.

Chapter 16 : Incubation, Fruiting Induction, Environmental Control, and Crop Management

"Growing *Cordyceps militaris* is not a matter of luck—it is the art of controlling the environment. The cultivator who masters temperature, humidity, light, and fresh air masters the harvest."

Learning Objectives

After completing this chapter, you will be able to:

- Understand the different stages of *Cordyceps militaris* development.
- Manage incubation and fruiting conditions effectively.
- Control temperature, humidity, lighting, carbon dioxide, and fresh air exchange.
- Identify the environmental factors affecting yield and quality.
- Monitor crop development using scientific methods.

- Prevent environmental stress and crop failure.

16.1 Introduction

Once inoculation has been completed successfully, the cultivation process enters its most critical stage—**crop management**. During this period, the fungus transitions from vegetative growth (mycelial colonization) to reproductive growth (fruiting body formation).

Unlike many edible mushrooms, *Cordyceps militaris* is highly responsive to subtle changes in environmental conditions. Even small deviations in temperature, humidity, or ventilation may affect:

- Mycelial growth
- Primordia initiation
- Fruiting body development
- Colour
- Yield
- Cordycepin content
- Overall product quality

Successful cultivation therefore depends on maintaining a stable environment throughout the production cycle.

16.2 Stages of Crop Development

The life cycle during cultivation may be divided into the following stages:

Stage 1 – Incubation

- Mycelium colonizes the substrate.
- White fungal growth spreads throughout the bottle.
- Little or no light is required.

Stage 2 – Complete Colonization

- Entire substrate becomes white.
- Mycelium becomes dense and vigorous.
- Bottle is ready for fruiting induction.

Stage 3 – Primordia Formation

Small pin-like structures begin to appear.

These represent the earliest stage of fruiting body development.

Stage 4 – Fruiting Body Elongation

The primordia gradually elongate.

Orange pigmentation becomes visible.

Stage 5 – Maturation

Fruiting bodies attain their final height, colour, and texture.

Harvest should be planned at the appropriate maturity stage.

16.3 Incubation Room Management

During incubation, the primary objective is rapid and uniform colonization.

Recommended Conditions

Parameter	**Recommended Range**
Temperature	20–24°C
Relative Humidity	Ambient laboratory conditions
Light	Darkness or very low light
Air Movement	Minimal
Bottle Disturbance	None

Maintain stable conditions and avoid unnecessary handling of bottles.

16.4 Fruiting Induction

Once complete colonization is achieved, bottles are transferred to the fruiting room.

Environmental conditions are gradually adjusted to stimulate the transition from vegetative growth to fruiting.

Key factors include:

- Light exposure
- Fresh air exchange
- Temperature stability
- Relative humidity

Sudden environmental changes should be avoided.

16.5 Temperature Management

Temperature strongly influences fungal metabolism.

Excessively High Temperature

May result in:

- Reduced growth
- Poor pigmentation
- Increased contamination risk
- Abnormal fruiting

Excessively Low Temperature

May lead to:

- Slow development
- Delayed harvest
- Reduced productivity

Maintain a consistent temperature appropriate for the selected strain and cultivation protocol.

16.6 Humidity Control

Relative humidity affects:

- Primordia initiation
- Fruiting body elongation

- Surface appearance
- Product quality

Low Humidity

Possible effects:

- Dry fruiting bodies
- Cracking
- Stunted growth

Excessively High Humidity

Possible effects:

- Condensation
- Increased contamination
- Soft tissues
- Poor air circulation

Humidity should be monitored continuously using calibrated hygrometers.

16.7 Light Management

Unlike green plants, *Cordyceps militaris* does not use light for photosynthesis.

However, light acts as a developmental signal.

Proper lighting encourages:

- Fruiting initiation
- Uniform growth
- Characteristic orange pigmentation

Commercial farms commonly use cool white LED lighting with a controlled photoperiod.

Avoid excessive heat from lighting systems.

16.8 Fresh Air Exchange (FAE)

Fresh air is essential during fruiting.

As mycelium grows, it consumes oxygen and releases carbon dioxide.

Without adequate ventilation:

- Fruiting bodies may become elongated and weak.
- Growth may become irregular.
- Crop quality may decline.

Proper ventilation maintains healthy gas balance while minimizing contamination.

16.9 Carbon Dioxide (CO_2) Management

High carbon dioxide concentrations can influence:

- Stem elongation
- Fruiting body morphology
- Yield
- Overall appearance

Commercial facilities often monitor CO_2 levels using electronic sensors.

For small-scale growers, regular fresh air exchange is generally sufficient when performed under clean conditions.

16.10 Air Circulation

Good air circulation helps:

- Maintain uniform temperature
- Reduce localized humidity
- Prevent condensation

- Improve environmental consistency

Avoid directing strong airflow directly onto the bottles, as this may dry developing fruiting bodies.

16.11 Water Management

Water should **not** be sprayed directly into cultivation bottles unless a validated cultivation method specifically requires it.

Instead:

- Maintain ambient humidity.
- Prevent standing water.
- Keep the cultivation room clean and dry.

Condensation inside bottles should be monitored, as excessive moisture may indicate environmental imbalance.

16.12 Daily Crop Monitoring

Inspect each batch daily.

Observe:

- Mycelial colour

- Primordia development
- Fruiting body height
- Bottle cleanliness
- Condensation
- Contamination
- Environmental readings

Use standardized observation sheets for consistency.

16.13 Crop Development Timeline

The exact duration varies with strain and cultivation conditions, but a typical production cycle may follow this sequence:

Stage	**Approximate Duration***
Incubation	2–3 weeks
Primordia Initiation	1 week
Fruiting Body Development	2–3 weeks
Harvest	5–7 weeks from inoculation

*The timeline is indicative and should be validated for the specific strain and production system.

16.14 Crop Monitoring Register

Maintain records of:

- Batch Number
- Date of Inoculation
- Colonization Status
- Primordia Appearance
- Fruiting Progress
- Harvest Date
- Yield
- Remarks

Consistent documentation allows comparison between production cycles.

16.15 Recognizing Healthy Fruiting Bodies

Healthy fruiting bodies generally exhibit:

- Bright orange colour
- Upright growth

- Uniform shape
- Firm texture
- Clean bottle surface
- No visible contamination

16.16 Common Environmental Problems

Problem: Poor Primordia Formation

Possible causes:

- Inadequate light
- Incorrect environmental conditions
- Weak culture

Problem: Thin Fruiting Bodies

Possible causes:

- Insufficient fresh air
- Strain characteristics
- Nutritional imbalance

Problem: Pale Colour

Possible causes:

- Inadequate lighting
- Premature harvest
- Suboptimal environmental management

Problem: Water Droplets Inside Bottle

Possible causes:

- Excess humidity
- Temperature fluctuations
- Poor ventilation

Problem: Uneven Growth

Possible causes:

- Inconsistent substrate preparation
- Uneven environmental conditions
- Variable inoculum quality

16.17 Environmental Monitoring Instruments

Every commercial cultivation room should include:

- Digital thermometer
- Hygrometer

- Timer for lighting
- CO_2 monitor (recommended)
- Air circulation fans
- Data logger (optional but valuable)

Regular calibration ensures accurate measurements.

Daily Crop Management Checklist

Before leaving the cultivation room each day:

☐ Record temperature.

☐ Record humidity.

☐ Check lighting system.

☐ Inspect ventilation.

☐ Observe crop development.

☐ Remove contaminated bottles promptly.

☐ Clean the room if necessary.

☐ Update production records.

Weekly Management Checklist

Every week:

✔ Inspect electrical equipment.

✔ Calibrate measuring instruments if scheduled.

✔ Review contamination rates.
✔ Compare growth between batches.
✔ Evaluate environmental trends.
✔ Plan harvesting schedules.

Troubleshooting Guide

Problem	**Possible Cause**	**Corrective Action**
Delayed fruiting	Environmental conditions not optimized	Review temperature, light, and ventilation settings
Abnormal morphology	Excess CO_2 or inadequate fresh air	Improve controlled ventilation
Surface contamination	Poor sanitation	Remove affected bottles and disinfect the room
Uneven pigmentation	Uneven lighting	Improve light distribution
Variable yield	Inconsistent substrate or spawn quality	Standardize preparation procedures

Best Practices for Commercial Growers

✔ Separate incubation and fruiting rooms.

✔ Install environmental monitoring systems.

✔ Train staff to recognize early signs of contamination.

✔ Use preventive maintenance for climate-control equipment.

✔ Analyze production records after every batch.

✔ Standardize operating procedures across all production rooms.

Chapter Summary

This chapter explained how environmental management influences every stage of *Cordyceps militaris* cultivation. You learned the principles of incubation, fruiting induction, temperature regulation, humidity control, lighting, fresh air exchange, and crop monitoring. Proper environmental control not only increases yield but also improves product quality and production consistency.

Key Takeaways

✓ Stable environmental conditions are essential for successful fruiting.

✓ Light, humidity, and fresh air must be carefully balanced.

✓ Daily monitoring prevents minor problems from becoming major crop losses.

✓ Accurate records enable continuous improvement in commercial production.

Chapter 17: Contamination

Identification, Prevention, Diagnosis, and Management

"In Cordyceps cultivation, contamination is not merely an inconvenience—it is the greatest threat to productivity. The most successful cultivators are not those who never encounter contamination, but those who identify it early, understand its source, and prevent it from recurring."

Learning Objectives

After completing this chapter, you will be able to:

- Understand the major causes of contamination.
- Identify bacterial, fungal, and yeast contaminants.
- Diagnose contamination at different cultivation stages.
- Implement preventive laboratory practices.
- Apply corrective measures to reduce crop

losses.

- Develop a contamination monitoring system for commercial production.

17.1 Introduction

Every mushroom cultivation laboratory faces contamination. Even the most advanced commercial farms occasionally lose production batches due to unwanted microorganisms.

Contamination may occur during:

- Media preparation
- Sterilization
- Cooling
- Inoculation
- Incubation
- Fruiting
- Harvesting
- Packaging

Understanding contamination is therefore one of the most valuable skills a Cordyceps cultivator can develop.

Early detection often saves an entire production room.

17.2 What is Contamination?

Contamination refers to the presence of unwanted microorganisms growing alongside *Cordyceps militaris.*

These unwanted organisms compete for:

- Nutrients
- Water
- Oxygen
- Space

Some contaminants also produce toxic metabolites that inhibit fungal growth.

17.3 Sources of Contamination

Contaminants may originate from many sources.

Raw Materials

- Poor-quality rice
- Contaminated silkworm pupae

- Dirty water
- Low-grade supplements

Air

- Dust
- Airborne spores
- Polluted ventilation
- Construction dust

Equipment

- Improperly cleaned bottles
- Dirty forceps
- Contaminated scalpels
- Poorly maintained laminar flow cabinet

Personnel

- Unwashed hands
- Dirty clothing
- Talking during inoculation
- Sneezing
- Touching sterile surfaces

Insects

- Fruit flies
- Houseflies
- Ants
- Cockroaches
- Stored-grain insects

Laboratory Environment

- Poor cleaning
- Standing water
- Dirty shelves
- Damaged walls
- Old contaminated cultures

17.4 Types of Contamination

The major categories include:

- Bacterial contamination
- Mould contamination
- Yeast contamination
- Mixed contamination

17.5 Bacterial Contamination

Bacteria multiply rapidly and often become visible before fungal growth.

Common Symptoms

- Wet substrate
- Sticky appearance
- Yellow slime
- Cloudy liquid
- Sour smell
- Soft substrate

Causes

- Insufficient sterilization
- Poor hygiene
- Excess moisture
- Contaminated inoculum
- Poor-quality water

Prevention

- Use proper sterilization.
- Maintain aseptic technique.

- Avoid excessive substrate moisture.
- Use healthy cultures.

17.6 Green Mould (*Trichoderma*)

One of the most destructive fungal contaminants. Initially appears white. Later becomes bright green due to spore production.

Characteristics

- Fast growth
- Aggressive competition
- Large numbers of spores
- Rapid spread

Common Causes

- Dirty inoculation room
- Contaminated air
- Poor laboratory hygiene
- Old cultures

Management

- Immediately isolate contaminated bottles.

- Do not open contaminated bottles inside the laboratory.
- Sterilize before disposal.
- Clean surrounding areas thoroughly.

17.7 Black Mould

Black moulds may appear as dark colonies on the substrate or bottle walls.

Possible causes include:

- Poor sterilization
- Air contamination
- Improper storage

These cultures should be discarded immediately.

17.8 Blue or Green-Blue Mould

Certain fungal contaminants produce bluish-green colonies. These should never be confused with healthy Cordyceps mycelium. Immediate disposal is recommended.

17.9 Pink Mould

Pink colonies indicate contamination. Possible sources include:

- Poor laboratory sanitation
- Contaminated water
- Improper handling

Never attempt to recover these cultures.

17.10 Yeast Contamination

Yeasts generally appear as:

- Cream-coloured
- Smooth
- Shiny
- Moist colonies

Although some yeasts grow slowly, they compete with the mycelium for nutrients.

17.11 Mixed Contamination

Occasionally more than one contaminant occurs simultaneously.

Example:

- Bacteria + Green Mould
- Yeast + Black Mould
- Mixed contamination usually indicates serious failures in laboratory hygiene.

17.12 Contamination During Agar Culture

Possible signs:

- Irregular colonies
- Multiple colony types
- Coloured patches
- Liquid accumulation
- Unpleasant odour
- Poor mycelial growth

17.13 Contamination During Grain Spawn

Symptoms include:

- Wet grains
- Green patches
- Black spots
- Yellow liquid
- Gas production

- Bottle swelling

17.14 Contamination During Fruiting

Visible symptoms:

- Abnormal colours
- Soft fruiting bodies
- Foul smell
- Water-soaked substrate
- Premature death of fruiting bodies

17.15 Environmental Causes

Environmental imbalance often increases contamination. Examples include:

- High humidity
- Poor ventilation
- Water condensation
- Dust accumulation
- Temperature fluctuations

17.16 Human Errors Leading to Contamination

Common mistakes include:

- Opening bottles unnecessarily
- Touching sterile tools
- Poor hand washing
- Talking while inoculating
- Working too quickly
- Ignoring contaminated cultures
- Using expired cultures
- Improper labeling

17.17 Diagnosis Flow Chart

Bottle Appears Abnormal

|

▼

Observe Colour

|

▼

White Only?

| |

Yes No

| |

Healthy Check Colour

|

┌──────────┼──────────
──┐
| | |

Green Black Yellow/Slime

| | |

Likely Mould Bacterial

Trichoderma Contamination

17.18 Laboratory Response Plan

When contamination is detected:

Step 1

Label affected bottles.

Step 2

Move immediately to quarantine area.

Step 3

Inspect nearby bottles.

Step 4

Record contamination.

Step 5

Sterilize contaminated bottles.

Step 6

Dispose safely.

Step 7

Disinfect laboratory.

Step 8

Investigate root cause.

17.19 Preventive Measures

Good Laboratory Practice (GLP)

✔ Daily cleaning

✔ Weekly deep cleaning

✔ UV sterilization (where appropriate)

✔ Regular filter maintenance

✔ Staff hygiene training

✔ Restricted laboratory access

✔ Pest control

✔ Routine environmental monitoring

17.20 Contamination Register

Maintain records including:

- Batch Number
- Date
- Culture Type
- Stage
- Type of Contamination
- Suspected Cause

- Corrective Action
- Operator Name

These records help identify recurring problems.

17.21 Commercial Quality Assurance

Every commercial laboratory should establish:

- Standard Operating Procedures
- Cleaning schedules
- Equipment maintenance records
- Environmental monitoring
- Training programmes
- Internal quality audits
- Corrective and preventive action (CAPA) procedures

Colour Identification Table

Colour Observed	Possible Interpretation	Recommended Action
White	Healthy mycelium (if uniform)	Continue monitoring
Bright Green	*Trichoderma* contamination	Isolate and sterilize
Black	Mould contamination	Discard safely
Yellow slime	Bacterial contamination	Discard and review sterilization
Pink	Fungal contamination	Remove immediately
Cream	Possible yeast contamination	Discard and investigate

Daily Inspection Checklist

- ☐ Bottle labels readable
- ☐ Mycelium uniformly white
- ☐ No coloured colonies
- ☐ No foul odour
- ☐ No condensation problems

☐ No damaged bottles
☐ Environmental parameters recorded
☐ Room cleaned

Troubleshooting Guide

Problem	Likely Cause	Corrective Action
Repeated bacterial contamination	Poor sterilization	Validate sterilization cycle
Green mould in many bottles	Contaminated inoculation room	Improve aseptic practices and clean thoroughly
High contamination after cooling	Bottles exposed too long	Reduce handling time before inoculation
Localized contamination	Equipment or operator issue	Inspect tools and retrain personnel
Increasing contamination over time	Poor laboratory sanitation	Conduct a comprehensive cleaning and review SOPs

Case Study 17.1 – A Commercial Laboratory Experience

A small commercial farm noticed that nearly 18% of its production bottles developed green mould within one week of inoculation. Investigation revealed that the pre-filter of the laminar airflow cabinet had not been cleaned for several months, reducing airflow efficiency and allowing dust accumulation. After replacing the pre-filter, servicing the HEPA filtration system, reinforcing aseptic techniques, and introducing a weekly sanitation schedule, contamination fell to below 3% over the next several production cycles.

Lesson: Contamination control depends on the entire production system—not on a single piece of equipment.

Practical Tips

✔ Never ignore a single contaminated bottle.

✔ Remove contaminated cultures immediately.

✔ Investigate every contamination incident.

✔ Keep contamination records permanently.

✔ Train every laboratory worker in aseptic techniques.

✔ Maintain separate clean and dirty areas.

✔ Review contamination statistics every month.

Chapter Summary

This chapter examined the major sources and types of contamination encountered during *Cordyceps militaris* cultivation. It explained how to recognize bacterial, fungal, and yeast contaminants, identify their likely causes, implement preventive measures, and establish systematic laboratory response procedures. Effective contamination management is essential for maintaining productivity, ensuring product quality, and reducing financial losses.

Key Takeaways

✓ Contamination is best prevented rather than treated.

✓ Early detection minimizes crop loss.

✓ Good Laboratory Practices (GLP) are the cornerstone of contamination control.

✓ Detailed records support continuous improvement and commercial quality assurance.

Chapter 18: Harvesting, Drying, Post-Harvest Handling, Packaging, Storage, and Quality Evaluation

"Growing a perfect crop is only half the journey. Harvesting at the right stage and preserving quality through proper post-harvest management determine the true commercial value of *Cordyceps militaris*."

Learning Objectives

After completing this chapter, you will be able to:

- Identify the ideal harvest stage of *Cordyceps militaris*.
- Harvest fruiting bodies without damaging the crop.
- Compare different drying methods.
- Understand post-harvest quality management.
- Select suitable packaging materials.
- Store dried Cordyceps properly.
- Grade products according to commercial

standards.

- Prepare products for domestic and export markets.

18.1 Introduction

The cultivation process does not end when the fruiting bodies mature. Improper harvesting or poor post-harvest handling can significantly reduce product quality, appearance, shelf life, and market value.

Commercial buyers evaluate Cordyceps based on:

- Colour
- Shape
- Size
- Cleanliness
- Moisture content
- Aroma
- Uniformity
- Packaging
- Traceability

A scientifically managed post-harvest system preserves both quality and customer confidence.

18.2 Determining the Correct Harvest Time

Harvesting too early may reduce biomass and overall yield, while harvesting too late can result in over-mature fruiting bodies, reduced visual appeal, and possible deterioration in quality.

Characteristics of an Ideal Harvest Stage

✔ Bright orange to deep orange colour

✔ Well-developed fruiting bodies

✔ Firm texture

✔ Uniform growth

✔ No visible contamination

✔ Clean bottle surface

Each production system should establish its own harvest criteria based on experience and quality objectives.

18.3 Harvesting Equipment

Use clean and sanitized equipment such as:

- Stainless steel forceps

- Sterile scissors
- Sharp scalpel
- Collection trays
- Gloves
- Face mask

Always disinfect tools before beginning a harvesting session.

18.4 Harvesting Procedure

Step 1

Inspect bottles individually.

Reject contaminated bottles immediately.

Step 2

Prepare a clean harvesting table.

Disinfect the working surface.

Step 3

Open one bottle at a time.

Avoid exposing other bottles unnecessarily.

Step 4

Remove the fruiting bodies carefully without introducing dirt or foreign material.

Step 5

Place harvested mushrooms gently into clean collection trays.

Avoid compression.

Step 6

Transfer immediately to the drying area.

Do not leave fresh mushrooms exposed for prolonged periods.

18.5 Harvest Hygiene

During harvesting:

✔ Wear gloves.

✔ Wear a mask.

✔ Avoid unnecessary talking.

✔ Keep fingernails short.

✔ Do not eat or drink in the harvesting room.

✔ Handle mushrooms gently.

18.6 Cleaning After Harvest

Generally, cultivated *Cordyceps militaris* requires minimal cleaning.

Remove only:

- Loose substrate particles
- Broken pieces
- Foreign materials

Avoid washing with water unless required by a validated processing method, as additional moisture increases drying time and may affect product quality.

18.7 Drying: Why It Is Important

Fresh Cordyceps contains a high proportion of water. Drying:

- Extends shelf life.
- Reduces microbial growth.
- Facilitates storage and transport.
- Helps preserve quality when performed correctly.

The drying method should minimize quality loss while producing a stable product.

18.8 Drying Methods

Natural Air Drying

Suitable only where environmental conditions are consistently clean and dry.

Advantages:

- Low cost
- No electricity required

Limitations:

- Slow
- Weather dependent
- Higher contamination risk

Hot-Air Drying

The most widely used commercial method.

Advantages:

- Controlled process
- Faster than natural drying

- Suitable for routine production

Care should be taken to avoid excessive temperatures that may reduce product quality.

Freeze-Drying (Lyophilization)

Considered a premium preservation method.

Advantages:

- Excellent shape retention
- Good preservation of colour
- Minimal structural damage

Limitations:

- High equipment cost
- Greater energy requirements

Often used for premium products and research applications.

Vacuum Drying

Provides controlled drying under reduced pressure.

Useful for selected high-value products where quality preservation is important.

18.9 Moisture Content

The dried product should reach a moisture level suitable for long-term storage and regulatory requirements applicable in the target market.

Excess moisture may lead to:

- Mould growth
- Product spoilage
- Reduced shelf life

Periodic moisture testing is recommended for commercial production.

18.10 Grading of Dried Cordyceps

Commercial producers often classify products into different grades.

Premium Grade

Characteristics:

- Uniform size

- Bright orange colour
- Straight fruiting bodies
- Clean appearance
- No damage
- Excellent packaging

Grade A

Minor natural variation in size.

Good colour.

Minimal defects.

Grade B

Moderate variation.

Some broken fruiting bodies.

Suitable for powder production or extraction.

Processing Grade

Broken or irregular material suitable for manufacturing capsules, powders, extracts, or functional foods.

18.11 Packaging Materials

Choose packaging that protects against:

- Moisture
- Oxygen
- Light
- Physical damage

Suitable materials include:

- Aluminium foil pouches
- Food-grade laminated pouches
- Glass jars with airtight lids
- High-barrier plastic containers

18.12 Vacuum Packaging

Vacuum packaging removes much of the air surrounding the product.

Benefits include:

✔ Reduced oxidation

✔ Better appearance

✔ Improved shelf life

✔ Reduced insect infestation

Vacuum packaging should be performed only after the product has been adequately dried.

18.13 Labeling Requirements

A commercial package should include:

- Product name
- Scientific name (*Cordyceps militaris*)
- Net weight
- Batch number
- Date of packing
- Best-before date
- Storage instructions
- Manufacturer details
- Country of origin (where required)

If sold as a food or dietary supplement, ensure compliance with the applicable labeling regulations in the target market.

18.14 Storage Conditions

Store dried products:

- In a cool place
- In a dry environment
- Away from direct sunlight
- Away from strong odours

- In sealed containers

Inspect stored products regularly for signs of moisture or infestation.

18.15 Shelf Life

Shelf life depends on:

- Drying efficiency
- Packaging quality
- Storage conditions
- Product moisture
- Handling practices

Commercial producers should establish shelf-life studies appropriate to their product and packaging system.

18.16 Quality Evaluation

Quality assessment may include:

Physical Parameters

- Colour
- Size

- Uniformity
- Texture
- Cleanliness

Chemical Parameters

Where laboratory facilities are available, evaluation may include:

- Moisture content
- Ash content
- Protein content
- Selected bioactive compounds (such as cordycepin)
- Heavy metal testing (as applicable)

Microbiological Quality

Commercial products should be tested according to applicable food safety standards for:

- Total microbial count
- Yeasts and moulds
- Indicator organisms, where required

Testing should be performed by competent laboratories.

18.17 Record Keeping

Maintain records of:

- Harvest date
- Batch number
- Drying method
- Drying duration
- Packaging date
- Storage location
- Product grade
- Final weight

These records support traceability and quality assurance.

Export Readiness Checklist

Before exporting, verify:

☐ Product fully dried

☐ Packaging intact

☐ Labels complete

☐ Batch records available

- ☐ Quality-control documents prepared
- ☐ Regulatory requirements reviewed
- ☐ Shipping cartons properly sealed

Commercial Packaging Options

Product Type	Recommended Package
Whole dried fruiting bodies	Vacuum-sealed aluminium pouch or glass jar
Powder	Moisture-barrier pouch or HDPE container
Capsules	Food-grade bottle with tamper-evident seal
Tea blend	Laminated tea sachets with outer pouch
Gift pack	Decorative rigid box with moisture-barrier inner pack

Common Post-Harvest Problems

Problem	Possible Cause	Corrective Action
Product softens during storage	Moisture absorption	Improve drying and use moisture-barrier packaging
Colour fading	Exposure to light or excessive heat	Store in opaque packaging away from sunlight
Mould growth	Inadequate drying or poor storage	Discard affected product and review drying procedures
Broken fruiting bodies	Rough handling	Improve harvesting and packaging practices
Insect infestation	Poor storage hygiene	Inspect storage regularly and maintain clean facilities

Practical Tips

✔ Harvest only healthy, contamination-free bottles.

✔ Dry the product promptly after harvest.

✔ Handle fruiting bodies gently to preserve appearance.

✔ Use food-grade packaging materials.

✔ Maintain batch traceability from cultivation to sale.

✔ Store samples from each production batch for quality reference.

✔ Periodically review customer feedback to improve product presentation.

Case Study 18.18 – Improving Product Value Through Better Packaging

A small producer sold dried *Cordyceps militaris* in simple polyethylene bags. Although the product quality was good, customers reported moisture problems during the rainy season. After switching to laminated aluminium pouches with vacuum sealing and adding clear batch labels, customer complaints declined significantly, product shelf stability improved, and the producer was able to command a higher selling price.

Lesson: Packaging is not merely a container—it is an important part of product quality and brand reputation.

Chapter Summary

This chapter described the complete post-harvest management process for *Cordyceps militaris*, from determining the correct harvest stage to drying, grading, packaging, storage, and quality evaluation. Proper post-harvest handling preserves the value created during cultivation and helps producers meet the expectations of consumers and commercial buyers.

Key Takeaways

✓ Harvest at the appropriate maturity stage.

✓ Dry the product promptly using a suitable method.

✓ Protect the product from moisture, light, and contamination.

✓ Maintain complete batch records for traceability.

✓ Attractive, informative packaging enhances commercial value.

Chapter 19 : Value Addition, Product Development, Branding, and Commercialization of Cordyceps militaris

"The greatest profit is not earned by selling mushrooms—it is earned by creating trusted products that solve consumer needs through quality, innovation, and scientific credibility."

Learning Objectives

After completing this chapter, you will be able to:

- Understand the concept of value addition.
- Develop commercially viable Cordyceps-based products.
- Learn basic processing methods.
- Understand branding and packaging strategies.
- Calculate production costs and pricing.
- Build a successful Cordyceps business.
- Understand quality, regulatory, and marketing considerations.

19.1 Introduction

Cultivating *Cordyceps militaris* is only the first step toward building a profitable enterprise. The real commercial potential lies in transforming harvested mushrooms into **high-value products** that meet consumer demand for convenience, quality, and innovation.

A grower selling fresh mushrooms earns one level of income, while an entrepreneur producing premium dried mushrooms, capsules, extracts, teas, or wellness products can often achieve significantly higher value from the same raw material.

Value addition also helps:

- Extend shelf life
- Improve product stability
- Reduce post-harvest losses
- Increase customer appeal
- Expand into new markets
- Build a recognizable brand

19.2 What is Value Addition?

Value addition means improving a raw agricultural or biological product through processing, packaging, branding, or formulation so that it commands a higher market value.

Examples include:

- Drying fresh mushrooms
- Producing mushroom powder
- Manufacturing capsules
- Preparing herbal tea blends
- Producing extracts
- Creating gift packs
- Developing nutraceutical products

19.3 Product Categories

Cordyceps products may be divided into several categories.

Primary Products

- Fresh fruiting bodies
- Dried fruiting bodies

Secondary Products

- Powder
- Capsules
- Tablets
- Tea
- Coffee blends

Advanced Products

- Standardized extracts
- Functional beverages
- Health supplements
- Nutraceutical formulations
- Cosmetic ingredients

19.4 Dried Whole Fruiting Bodies

The premium commercial product remains the whole dried fruiting body.

Advantages:

✔ Premium appearance

✔ Minimal processing

✔ Longer shelf life

✔ High consumer confidence

Ideal customers include:

- Herbal stores
- Wellness retailers
- Exporters
- Online buyers

19.5 Cordyceps Powder

Powder is produced by grinding properly dried fruiting bodies under hygienic conditions.

Applications:

- Capsules
- Functional foods
- Smoothies
- Tea blends
- Health drinks

Grinding equipment should be thoroughly cleaned before use to prevent cross-contamination.

19.6 Capsule Manufacturing

Powdered Cordyceps can be filled into hard-shell capsules using manual, semi-automatic, or automatic capsule-filling machines.

Advantages

- Convenient dosage
- Easy transportation
- Attractive retail presentation
- Improved consumer acceptance

Each production batch should be tested for weight uniformity and filled under hygienic conditions.

19.7 Tablet Production

Tablet manufacturing generally requires specialized equipment and suitable excipients. Commercial tablet production should follow established pharmaceutical or nutraceutical manufacturing standards, as applicable.

19.8 Herbal Tea Blends

Cordyceps powder or small dried pieces may be blended with herbs such as:

- Green tea
- Lemongrass
- Ginger
- Tulsi (*Ocimum tenuiflorum*)
- Peppermint
- Cinnamon

Blend formulations should be developed with attention to flavour, aroma, and regulatory requirements.

19.9 Cordyceps Coffee

Functional coffee blends have become increasingly popular.

Typical ingredients may include:

- Instant coffee
- Cordyceps powder
- Chicory
- Natural flavouring

Manufacturers should clearly label ingredients and avoid unsupported health claims.

19.10 Functional Foods

Cordyceps may be incorporated into selected foods, including:

- Protein bars
- Biscuits
- Nutrition mixes
- Breakfast cereals
- Soup powders

Product stability and sensory quality should be evaluated during development.

19.11 Beverage Development

Commercial opportunities include:

- Herbal drinks
- Wellness beverages
- Energy formulations
- Ready-to-drink products

Each formulation should undergo stability testing and comply with applicable food regulations.

19.12 Extract Preparation

Extracts are prepared using controlled extraction methods with suitable solvents, depending on the target compounds and intended use.

Commercial extraction should be standardized to ensure batch-to-batch consistency.

Where claims are based on specific constituents such as cordycepin, appropriate analytical testing should be performed.

19.13 Cosmetic Applications

Research into mushroom-derived cosmetic ingredients is expanding.

Potential applications include:

- Face serums
- Skin creams
- Facial masks

- Storage instructions
- Manufacturer details
- Customer support information

Avoid making disease-treatment or cure claims unless specifically permitted under applicable regulations.

19.17 Pricing Strategy

Pricing depends upon:

- Production cost
- Packaging cost
- Labour
- Marketing expenses
- Distribution
- Profit margin
- Market positioning

Premium products generally justify premium packaging and customer support.

Example Cost Analysis

Expense	**Percentage of Total Cost (Illustrative)**
Raw materials	30%
Labour	20%
Packaging	15%
Utilities	10%
Marketing	10%
Distribution	5%
Administrative expenses	5%
Contingency	5%

Actual costs will vary depending on production scale and location.

19.18 Marketing Channels

Possible sales channels include:

Direct Sales

- Farm outlet
- Retail shop
- Wellness centre

Online Sales

- Company website
- E-commerce platforms
- Social media stores

Business-to-Business (B2B)

- Pharmacies
- Nutraceutical companies
- Herbal product manufacturers
- Health-food distributors

Export

International markets often require additional documentation, quality assurance, and compliance with import regulations.

19.19 Customer Education

Consumers appreciate accurate information. Provide educational material explaining:

- Product origin
- Cultivation method
- Storage instructions

- Recommended usage
- Quality-control practices

Well-informed customers are more likely to become repeat buyers.

19.20 Good Manufacturing Practices (GMP)

Value-added products should be produced under Good Manufacturing Practices.

Key elements include:

- Hygienic production areas
- Trained personnel
- Equipment sanitation
- Batch documentation
- Traceability
- Complaint handling
- Product recall procedures

19.21 Regulatory Considerations

Before marketing products, producers should review the regulatory requirements applicable in their country.

Depending on the product category, this may include:

- Food safety registration or licensing
- Labelling compliance
- Packaging standards
- Advertising regulations
- Quality testing requirements

Always ensure that product claims are supported by evidence and comply with applicable laws.

19.22 Building a Successful Cordyceps Brand

Successful companies generally focus on:

✔ Consistent quality

✔ Scientific transparency

✔ Attractive packaging

✔ Reliable customer service

✔ Honest communication

✔ Continuous product improvement

✔ Strong online presence

Business Growth Roadmap

Small Laboratory

|

▼

Fresh Mushroom Sales

|

▼

Dried Mushroom Production

|

▼

Premium Packaging

|

▼

Online Sales

|

▼

Value-Added Products

|

▼

National Distribution

|

▼

Export Opportunities

Business Checklist

Before launching a product:

☐ Product formulation finalized

☐ Packaging tested

☐ Labels verified

☐ Batch records prepared

☐ Product specifications documented

☐ Quality-control procedures established

☐ Customer service plan developed

☐ Marketing strategy prepared

Case Study 19.1 – From Small Grower to Premium Brand

A home-based cultivator began by selling small quantities of fresh *Cordyceps militaris* to local wellness enthusiasts. Within two years, the business expanded to produce dried fruiting bodies packaged in moisture-resistant pouches with professionally designed labels and batch numbers. The company later introduced mushroom powder and herbal tea blends while maintaining consistent quality and transparent product information. As customer confidence increased, online sales exceeded local retail sales, demonstrating that branding, packaging, and quality assurance can substantially increase the commercial value of a specialty mushroom enterprise.

Lesson: Long-term business success depends on trust, consistency, and quality—not simply on production volume.

Chapter Summary

This chapter explored the transformation of *Cordyceps militaris* from a cultivated mushroom into a range of value-added commercial products. It covered product development, branding, packaging, pricing, marketing channels, Good Manufacturing Practices, and business growth strategies. Entrepreneurs who combine scientific cultivation with professional product development can create sustainable businesses serving both domestic and international markets.

Key Takeaways

✓ Value addition significantly increases commercial potential.

✓ Professional branding builds customer confidence.

✓ Packaging should protect the product and communicate quality.

✓ Regulatory compliance is essential for long-term business success.

✓ Continuous innovation helps businesses remain competitive.

Chapter 20: Economics, Business Planning, Commercial Production Models, and Financial Analysis for Cordyceps militaris

"Scientific cultivation produces mushrooms. Scientific business planning produces profits. Every successful Cordyceps entrepreneur understands both biology and business."

Learning Objectives

After completing this chapter, you will be able to:

- Prepare a business plan for a Cordyceps cultivation unit.
- Estimate capital investment and operational costs.
- Understand different production scales.
- Calculate production costs and profitability.
- Evaluate Return on Investment (ROI).
- Identify risks and mitigation strategies.
- Explore funding opportunities.
- Develop a roadmap for business expansion.

20.1 Introduction

The commercial cultivation of *Cordyceps militaris* has evolved into a high-value biotechnology enterprise. With increasing consumer interest in functional mushrooms, nutraceuticals, and wellness products, entrepreneurs now have opportunities to establish profitable ventures ranging from small home laboratories to fully automated production facilities.

However, commercial success requires more than biological knowledge. Entrepreneurs must understand:

- Investment planning
- Production economics
- Cost control
- Marketing
- Risk management
- Quality assurance
- Regulatory compliance

A well-prepared business plan significantly increases the likelihood of long-term success.

20.2 Selecting the Right Business Model

The choice of business model depends on:

- Available capital
- Technical expertise
- Infrastructure
- Market access
- Production objectives

The following models are commonly adopted.

Model I – Home-Based Laboratory

Suitable for:

- Beginners
- Students
- Researchers
- Hobby growers

Production Capacity:
100–300 bottles per production cycle

Advantages:

- Low investment
- Easy management
- Minimal manpower
- Opportunity for skill development

Limitations:

- Limited production
- Lower income potential
- Restricted automation

Model II – Small Commercial Unit

Production Capacity:

500–2,000 bottles per cycle

Suitable for:

- Start-up entrepreneurs
- Self-Help Groups (SHGs)
- Farmer Producer Organizations (FPOs)
- Small businesses

Advantages:

- Better profitability
- Improved equipment utilization
- Local market supply

Model III – Medium Commercial Unit

Production Capacity:

2,000–10,000 bottles

Characteristics:

- Dedicated laboratory
- Separate incubation and fruiting rooms
- Climate-controlled production
- Quality-control laboratory

Suitable for:

- MSMEs
- Biotechnology start-ups
- Commercial mushroom farms

Model IV – Industrial Production Facility

Production Capacity:

More than 10,000 bottles per cycle

Facilities generally include:

- Clean rooms
- Automated filling lines
- Commercial autoclaves
- Climate-controlled cultivation rooms
- Freeze dryers
- Packaging units
- Quality-control laboratories

20.3 Infrastructure Requirements

A well-designed facility typically includes:

Room	Purpose
Raw Material Store	Storage of grains and supplements
Media Preparation Room	Preparation of substrate
Sterilization Room	Autoclave operation
Inoculation Room	Aseptic culture transfer
Incubation Room	Mycelial growth
Fruiting Room	Fruiting body production
Drying Room	Post-harvest processing
Packaging Room	Packing and labeling
Finished Goods Store	Product storage
Office	Administration

Proper workflow minimizes contamination risk.

20.4 Equipment Requirements

Typical equipment includes:

- Laminar Air Flow Cabinet
- Autoclave or Pressure Sterilizer

- Analytical Balance
- pH Meter
- Incubator
- Refrigerator
- Air Conditioner
- Hygrometer
- Thermometer
- HEPA Air Filters
- LED Lighting
- Bottle Washer
- Drying Machine
- Vacuum Packaging Machine
- Capsule Filling Machine (for value-added products)

20.5 Human Resource Requirements

For a medium-sized unit:

Position	Responsibilities
Production Manager	Overall supervision
Laboratory Technician	Culture preparation
Production Assistants	Bottle preparation and inoculation
Quality Control Officer	Testing and documentation
Packaging Staff	Packing and labeling
Marketing Executive	Sales and customer support

Initially, several roles may be performed by the same individual in a small enterprise.

20.6 Capital Investment

Capital expenditure generally includes:

- Building or room modification
- Laboratory furniture
- Cultivation racks
- Laboratory equipment
- Air-conditioning

- Electrical installation
- Water purification system
- Drying equipment
- Packaging equipment

The exact investment depends on the production scale, automation level, and local costs.

20.7 Working Capital

Working capital covers recurring expenses such as:

- Raw materials
- Bottles
- Culture media
- Electricity
- Water
- Packaging materials
- Salaries
- Transportation
- Maintenance
- Marketing

Adequate working capital is essential to maintain uninterrupted production.

20.8 Production Cost Components

Typical recurring costs include:

Expense	**Approximate Share (%)**
Raw Materials	30
Labour	20
Utilities	10
Packaging	15
Marketing	10
Maintenance	5
Administration	5
Miscellaneous	5

These percentages are illustrative and should be adjusted based on local conditions.

20.9 Cost of Production

Production cost per bottle depends on:

- Substrate cost
- Spawn cost
- Labour

- Electricity
- Bottle depreciation
- Packaging
- Quality testing

Maintaining detailed production records enables accurate cost calculation.

20.10 Revenue Sources

A Cordyceps enterprise may generate income through:

- Fresh mushrooms
- Dried mushrooms
- Mushroom powder
- Capsules
- Extracts
- Tea blends
- Training programmes
- Consultancy services
- Spawn sales
- Tissue culture services

Diversified income streams improve business resilience.

20.11 Break-Even Analysis

The break-even point is reached when total income equals total expenditure.

Understanding the break-even point helps entrepreneurs determine:

- Minimum production required
- Minimum sales target
- Profitability timeline

Break-even analysis should be updated periodically as costs change.

20.12 Return on Investment (ROI)

ROI measures the efficiency of capital investment.

A higher ROI indicates better utilization of invested resources.

ROI should be evaluated over several production cycles rather than a single batch.

- Personal investment
- Bank loans
- Government entrepreneurship schemes
- MSME support programmes
- Startup incubators
- Venture capital (for large-scale enterprises)
- Cooperative financing

Applicants should prepare a detailed project report before seeking finance.

20.17 Marketing Strategy

Successful marketing combines:

- Product quality
- Attractive packaging
- Educational content
- Digital marketing
- Customer engagement

Useful marketing channels include:

- Company website
- Social media

- E-commerce platforms
- Health and wellness exhibitions
- Distributor networks

20.18 SWOT Analysis

Strengths	**Weaknesses**
High-value crop	Requires technical expertise
Growing demand	Initial capital investment
Premium pricing	Strict contamination control
Opportunities	**Threats**
Export markets	New competitors
Functional foods	Regulatory changes
Online sales	Counterfeit products

20.19 Standard Business Workflow

Market Research

|

▼

Business Plan

|
▼

Finance Arrangement

|
▼

Infrastructure Setup

|
▼

Laboratory Establishment

|
▼

Pilot Production

|
▼

Quality Improvement

|
▼

Commercial Production

|
▼

Brand Development

|

▼

Business Expansion

20.20 Record Keeping

Maintain separate records for:

- Production
- Sales
- Purchases
- Inventory
- Staff
- Equipment maintenance
- Customer complaints
- Financial transactions

Good documentation supports informed decision-making and regulatory compliance.

Business Planning Checklist

Before starting production:

☐ Market survey completed

☐ Business plan prepared

☐ Financial resources secured

☐ Infrastructure ready

☐ Equipment installed
☐ SOPs prepared
☐ Staff trained
☐ Trial production completed
☐ Marketing plan finalized
☐ Regulatory requirements reviewed

Case Study 20.1 – Building a Sustainable Enterprise

A biotechnology graduate established a small laboratory with 300 bottles per cycle. Rather than investing heavily in expensive automation during the first year, the entrepreneur focused on producing high-quality dried fruiting bodies, maintaining meticulous production records, and building a reputation through educational workshops and direct online sales. As demand increased, profits were reinvested in climate-controlled fruiting rooms and improved packaging equipment. Within five years, the business had expanded into capsule manufacturing and value-added products while maintaining consistent quality.

Lesson: Sustainable growth is often achieved through gradual expansion, continuous quality improvement, and reinvestment rather than rapid, high-risk scaling.

Practical Tips

✔ Start with a manageable production scale.

✔ Keep detailed financial records.

✔ Reinvest a portion of profits into quality improvement.

✔ Build a recognizable brand.

✔ Diversify products to reduce business risk.

✔ Monitor production costs every month.

✔ Invest in staff training and quality assurance.

Chapter Summary

This chapter presented the economic and business aspects of *Cordyceps militaris* cultivation. It discussed production models, infrastructure, equipment, human resources, investment planning, cost analysis, marketing, funding, risk management, and business expansion. A successful Cordyceps enterprise requires both scientific cultivation practices and sound business management.

Key Takeaways

✓ A detailed business plan is the foundation of commercial success.

✓ Production costs should be monitored continuously.

✓ Diversified income sources strengthen financial stability.

✓ Quality, branding, and customer trust are long-term competitive advantages.

✓ Sustainable growth is achieved through careful planning and disciplined management.

Chapter 21: Quality Assurance (QA), Quality Control (QC), Good Laboratory Practices (GLP), Good Manufacturing Practices (GMP), Documentation, and Regulatory Compliance

"A successful Cordyceps enterprise is built not only on high yields but also on consistency, traceability, safety, and customer confidence. Quality is never an accident—it is the result of a well-designed management system."

Learning Objectives

After completing this chapter, you will be able to:

- Understand the concepts of Quality Assurance (QA) and Quality Control (QC).
- Implement Good Laboratory Practices (GLP) and Good Manufacturing Practices (GMP).
- Develop Standard Operating Procedures (SOPs).
- Maintain comprehensive production records.

- Establish traceability and documentation systems.
- Understand regulatory requirements for commercial production.
- Prepare for inspections and quality audits.

21.1 Introduction

As the global demand for medicinal mushrooms grows, customers, distributors, and regulatory authorities increasingly expect products that are safe, consistent, and scientifically documented.

Commercial success is therefore closely linked to:

- Product quality
- Food safety
- Laboratory discipline
- Accurate documentation
- Regulatory compliance

Quality management is no longer optional—it is an essential part of modern mushroom cultivation.

21.2 What is Quality?

Quality refers to the degree to which a product consistently meets predetermined specifications and customer expectations.

For *Cordyceps militaris*, quality includes:

- Correct species identity
- Uniform appearance
- Freedom from contamination
- Appropriate moisture level
- Proper packaging
- Accurate labeling
- Batch traceability

21.3 Quality Assurance (QA)

Quality Assurance is a preventive system designed to ensure that every stage of production follows established procedures.

QA focuses on:

- Process control
- Staff training
- SOP implementation
- Documentation
- Preventive action
- Continuous improvement

QA aims to prevent defects before they occur.

21.4 Quality Control (QC)

Quality Control involves testing and inspecting products to verify that they meet defined quality standards.

Typical QC activities include:

- Raw material inspection
- Culture examination
- Sterility checks
- Environmental monitoring
- Moisture testing
- Packaging inspection
- Finished product evaluation

QC identifies problems before products reach customers.

21.5 Difference Between QA and QC

Quality Assurance (QA)	Quality Control (QC)
Preventive	Detective
Process-oriented	Product-oriented
Focuses on systems	Focuses on testing
Prevents errors	Detects defects
Continuous	Batch-specific

Both systems complement each other.

21.6 Good Laboratory Practices (GLP)

GLP provides guidelines for maintaining reliable laboratory operations.

Key principles include:

- Clean workspaces
- Proper equipment maintenance
- Staff competency
- Accurate record keeping
- Sample traceability
- Safe waste disposal

21.7 Good Manufacturing Practices (GMP)

GMP ensures that products are consistently produced under hygienic and controlled conditions.

Major components include:

Personnel Hygiene

- Clean uniforms
- Hair covers
- Gloves
- Face masks
- Hand washing

Facility Hygiene

- Clean floors
- Dust-free walls
- Controlled access
- Pest control
- Adequate lighting

Equipment Maintenance

- Routine servicing

- Calibration
- Cleaning schedules
- Maintenance logs

Process Control

- Standardized procedures
- Batch monitoring
- Documentation
- Verification

21.8 Standard Operating Procedures (SOPs)

Every commercial laboratory should prepare written SOPs.

Typical SOPs include:

- Culture preparation
- Agar media preparation
- Bottle filling
- Sterilization
- Inoculation
- Incubation
- Fruiting management
- Harvesting

- Drying
- Packaging
- Cleaning
- Waste disposal

All employees should receive training before performing these procedures independently.

21.9 Documentation

Good documentation provides evidence that procedures were followed correctly.

Typical records include:

- Batch Manufacturing Records (BMR)
- Batch Packaging Records (BPR)
- Equipment Logs
- Cleaning Records
- Temperature Logs
- Humidity Logs
- Culture Transfer Records
- Calibration Records
- Training Records

21.10 Batch Manufacturing Record (BMR)

Each production batch should include:

- Batch Number
- Date
- Strain Used
- Raw Materials
- Operator Name
- Sterilization Details
- Incubation Start Date
- Fruiting Date
- Harvest Date
- Final Yield

Proper batch records facilitate traceability.

Sample Batch Manufacturing Record

Item	Details
Batch No.	________
Date	________
Culture Code	________
Operator	________
Number of Bottles	________
Sterilization Cycle	________
Harvest Date	________
Yield	________

21.11 Equipment Logbook

Maintain records for:

- Autoclaves
- Laminar Air Flow Cabinets
- Dryers
- Air Conditioners
- Refrigerators
- Incubators

Each entry should include:

- Date
- Equipment ID
- Maintenance performed
- Technician
- Next due date

21.12 Environmental Monitoring

Record environmental conditions daily.

Typical parameters include:

Parameter	Monitoring Frequency
Temperature	Daily
Humidity	Daily
Lighting	Daily
CO_2 (if monitored)	Daily or as required
Airflow	Weekly
Cleanliness	Daily

21.13 Calibration

Instruments should be calibrated periodically.

Equipment requiring calibration includes:

- Thermometers

- Hygrometers
- Weighing balances
- pH meters
- Temperature controllers

Calibration records should be retained.

21.14 Traceability System

Every finished product should be traceable back to:

- Raw materials
- Culture batch
- Production records
- Harvest date
- Packaging date
- Operator
- Laboratory

This enables efficient investigation of quality issues.

21.15 Internal Audits

Periodic internal audits help verify compliance with SOPs.

Audit topics include:

- Laboratory cleanliness
- Documentation
- Equipment maintenance
- Personnel hygiene
- Production records
- Product labeling

Audit findings should be documented and corrective actions assigned.

21.16 Corrective and Preventive Action (CAPA)

CAPA is a structured method for resolving quality issues.

Corrective Action

Addresses the immediate problem.

Example:

Replacing contaminated media.

Preventive Action

Addresses the root cause.

Example:

Revising sterilization procedures and retraining staff.

21.17 Product Recall Procedure

Every commercial producer should establish a written recall procedure.

Typical steps include:

1. Identify affected batches.
2. Inform distributors.
3. Stop further distribution.
4. Retrieve affected products where appropriate.
5. Investigate the cause.
6. Implement CAPA.
7. Document the entire process.

21.18 Regulatory Considerations (India)

Entrepreneurs should familiarize themselves with applicable regulations before marketing products.

Depending on the product category, this may include:

- Food business registration or licensing.
- Packaging and labeling requirements.
- GST compliance.
- Trademark registration.
- Import/export documentation (where applicable).

Regulatory requirements may change over time, so always verify the latest official guidance.

21.19 Quality Certifications

As the business grows, certification may strengthen market confidence.

Examples include:

- ISO 9001 (Quality Management)
- ISO 22000 (Food Safety Management)

- HACCP-based food safety systems
- GMP certification (where applicable)
- Organic certification (if eligibility criteria are met)

Certification should be pursued only after establishing a stable quality management system.

Quality Management Flowchart

Raw Material Inspection
|
▼
Media Preparation
|
▼
Sterilization Verification
|
▼
Culture Quality Check
|
▼
Production Monitoring

|
▼
Harvest Inspection
|
▼
Drying Verification
|
▼
Packaging Inspection
|
▼
Finished Product Release

Daily QA Checklist

☐ Laboratory cleaned
☐ Equipment functioning
☐ Environmental conditions recorded
☐ SOP followed
☐ Culture health verified
☐ Contamination checked
☐ Documentation completed
☐ Waste disposed safely

Monthly QA Checklist

✔ Review contamination statistics
✔ Verify calibration schedule
✔ Audit documentation
✔ Review customer complaints
✔ Inspect storage conditions
✔ Conduct staff refresher training
✔ Update SOPs if necessary

Common Documentation Errors

Error	Consequence	Prevention
Missing batch number	Poor traceability	Use standardized labels
Incomplete records	Difficult investigations	Complete records immediately
Illegible handwriting	Misinterpretation	Use block letters or digital systems
Unsigned forms	Lack of accountability	Require operator signatures
Delayed documentation	Data inaccuracies	Record information in real time

Case Study 21.1 – Improving Traceability

A commercial producer received a customer complaint regarding moisture in one batch of dried *Cordyceps militaris*. Because each package carried a batch number linked to production records, the company traced the issue to a single drying cycle in which equipment performance had declined. The affected batch was isolated, corrective maintenance was completed, and the drying SOP was updated. No additional batches were affected.

Lesson: Effective documentation enables rapid problem identification, protects customers, and strengthens business credibility.

Practical Tips

✔ Develop simple, easy-to-follow SOPs.

✔ Document every important activity.

✔ Conduct routine internal audits.

✔ Train staff regularly.

✔ Keep quality records secure and organized.

✔ Review customer feedback as part of continuous improvement.

✔ Remember: "If it is not documented, it is difficult to verify."

Chapter Summary

This chapter introduced the principles of Quality Assurance, Quality Control, Good

Laboratory Practices, Good Manufacturing Practices, documentation, traceability, and regulatory compliance. These systems provide the foundation for producing safe, reliable, and commercially successful *Cordyceps militaris* products. By integrating quality into every stage of production, cultivators can reduce errors, build customer confidence, and support long-term business growth.

Key Takeaways

✓ Quality should be built into the process—not inspected in at the end.

✓ QA prevents problems; QC detects them.

✓ Documentation and traceability are essential for commercial production.

✓ Internal audits and CAPA support continuous improvement.

✓ Compliance with applicable regulations enhances market access and consumer trust.

Chapter 22: Scientific Research, Bioactive Compounds, Pharmacology, Clinical Evidence, and Future Prospects of Cordyceps militaris

*"Cordyceps militaris is no longer viewed merely as a medicinal mushroom. It has become an important subject of modern biotechnology, pharmacology, functional food research, and precision medicine. The future of Cordyceps lies at the intersection of traditional knowledge and evidence-based science."**

Learning Objectives

After completing this chapter, you will be able to:

- Understand the major bioactive compounds present in *Cordyceps militaris*.
- Learn their biological significance.
- Understand current pharmacological research.
- Differentiate laboratory findings from clinical evidence.
- Identify limitations of existing research.

- Explore future opportunities in biotechnology and pharmaceutical development.

22.1 Introduction

During the last three decades, *Cordyceps militaris* has attracted worldwide scientific attention due to its diverse bioactive compounds and its potential applications in health, nutrition, biotechnology, and pharmaceutical research.

Unlike the wild caterpillar fungus (*Ophiocordyceps sinensis*), *Cordyceps militaris* can be cultivated under controlled laboratory conditions, making it a sustainable and economically viable source of valuable natural compounds.

Research institutions across Asia, Europe, and North America continue to investigate its chemistry, cultivation, genomics, metabolomics, and potential biological activities.

It is important to recognize that while laboratory and animal studies are promising, many proposed health applications require further high-quality human clinical research.

22.2 Bioactive Compounds in *Cordyceps militaris*

The medicinal value of *Cordyceps militaris* is primarily attributed to its rich profile of naturally occurring bioactive compounds.

Major compounds include:

- Cordycepin (3'-deoxyadenosine)
- Adenosine
- Polysaccharides
- Ergosterol
- Ergothioneine
- Mannitol (Cordycepic acid)
- Carotenoids
- Peptides
- Proteins
- Phenolic compounds
- Flavonoid-like compounds
- Essential amino acids
- Vitamins
- Minerals

The concentration of these compounds depends on:

- Strain
- Substrate composition
- Cultivation conditions
- Harvest stage
- Drying method
- Storage conditions

22.3 Cordycepin

Cordycepin is considered the signature bioactive compound of *Cordyceps militaris*.

Chemically, it is a naturally occurring nucleoside analogue structurally related to adenosine.

Scientific Importance

Cordycepin has been widely investigated for its potential:

- Antioxidant activity
- Anti-inflammatory activity

- Immunomodulatory effects
- Antimicrobial activity
- Cellular signaling effects

Numerous laboratory studies suggest biological activity; however, the effectiveness and safety for specific medical uses in humans remain under investigation.

22.4 Adenosine

Adenosine is another naturally occurring nucleoside present in *Cordyceps militaris*.

It plays important physiological roles in living organisms and is often measured during quality evaluation of cultivated Cordyceps.

22.5 Polysaccharides

Polysaccharides are complex carbohydrates found in fungal cell walls.

Research suggests they may contribute to:

- Immune system modulation
- Antioxidant activity
- Functional food applications

Their structure varies depending on the fungal strain and cultivation conditions.

22.6 Ergosterol

Ergosterol is the principal sterol found in fungi.

It serves as an indicator of fungal biomass and is of scientific interest because it can be converted to vitamin D_2 under ultraviolet light.

22.7 Ergothioneine

Ergothioneine is a naturally occurring sulfur-containing amino acid derivative.

Current research is exploring its role as an antioxidant and its distribution in edible mushrooms, including *Cordyceps militaris*.

22.8 Mannitol (Cordycepic Acid)

Historically referred to as "cordycepic acid," mannitol contributes to the biochemical profile of *Cordyceps* species.

It is widely distributed in nature and is commonly detected in mushroom analyses.

22.9 Carotenoids

The characteristic orange colour of *Cordyceps militaris* is associated with carotenoid pigments.

These pigments:

- Influence product appearance.
- Are important quality indicators.
- May contribute antioxidant properties.

Light intensity and cultivation conditions influence pigment production.

22.10 Proteins and Amino Acids

Cordyceps militaris contains proteins and a range of amino acids.

Protein composition may vary depending on:

- Growth substrate
- Cultivation method
- Developmental stage

22.11 Vitamins and Minerals

Analytical studies have identified various vitamins and minerals in cultivated fruiting bodies.

Their concentrations depend on cultivation conditions and substrate composition.

22.12 Modern Analytical Techniques

Researchers use sophisticated analytical methods to identify and quantify bioactive compounds.

Common techniques include:

- High-Performance Liquid Chromatography (HPLC)
- Ultra-High Performance Liquid Chromatography (UHPLC)
- Liquid Chromatography-Mass Spectrometry (LC-MS)
- Gas Chromatography-Mass Spectrometry (GC-MS)
- Nuclear Magnetic Resonance (NMR)
- Fourier Transform Infrared Spectroscopy

(FTIR)

These techniques support quality control, research, and product standardization.

22.13 Pharmacological Research

Experimental studies have explored several biological activities of *Cordyceps militaris* extracts and isolated compounds.

Research areas include:

- Antioxidant mechanisms
- Immune modulation
- Anti-inflammatory responses
- Cellular metabolism
- Exercise physiology
- Neurobiology
- Metabolic regulation

Most of these findings come from laboratory experiments or animal studies and should not be interpreted as proof of clinical efficacy in humans.

22.14 Clinical Research

Human studies on *Cordyceps militaris* remain limited compared with laboratory research.

Existing clinical investigations have examined topics such as:

- Exercise performance
- Fatigue
- General wellness
- Metabolic parameters

Results have been mixed, and further well-designed, large-scale clinical trials are needed before definitive conclusions can be drawn.

22.15 Safety Considerations

Cultivated *Cordyceps militaris* has been consumed traditionally and is used in various food and supplement products.

However, consumers should note that:

- Supplements are not substitutes for prescribed medical treatment.

- Individuals with underlying medical conditions should consult healthcare professionals before use.
- Pregnant or breastfeeding women, children, and individuals taking medications should seek professional advice before using dietary supplements.

Manufacturers should ensure products are free from contaminants and comply with applicable food safety standards.

22.16 Biotechnology Applications

Modern biotechnology is expanding the applications of *Cordyceps militaris*.

Current areas of research include:

- Strain improvement
- Tissue culture
- Molecular breeding
- Genome sequencing
- Fermentation technology
- Metabolic engineering

- Bioactive compound enhancement

These advances may improve productivity and consistency in future commercial cultivation.

22.17 Genomics and Molecular Biology

Genome sequencing has provided insights into:

- Secondary metabolite pathways
- Fruiting body development
- Pigment synthesis
- Stress response mechanisms
- Gene regulation

Such knowledge supports future breeding and biotechnological innovation.

22.18 Artificial Intelligence in Cordyceps Research

Artificial Intelligence (AI) is beginning to assist researchers by:

- Predicting optimal cultivation conditions
- Monitoring environmental parameters

- Detecting contamination from images
- Analyzing production data
- Supporting quality control

Integration of AI with automated cultivation systems may improve efficiency and reduce production losses.

22.19 Future Research Directions

Promising areas include:

- Improved cultivation substrates
- High-cordycepin strains
- Precision fermentation
- Sustainable production systems
- Functional food development
- Personalized nutrition
- Controlled-environment agriculture
- Omics-based breeding
- Clinical validation

These fields are expected to shape the next generation of Cordyceps research.

Research Challenges

Current challenges include:

- Lack of standardized cultivation protocols
- Variation between strains
- Differences in analytical methods
- Limited multicenter clinical trials
- Regulatory differences among countries

Addressing these challenges will improve reproducibility and global acceptance.

Suggested Research Topics for Students

- Optimization of substrate formulations
- Effect of LED light spectra on pigment production
- Comparative study of drying methods
- Quantification of cordycepin under different cultivation conditions
- Evaluation of environmental stress responses
- AI-based contamination detection
- Shelf-life studies of dried products

- Packaging optimization
- Consumer acceptance studies

Emerging Technologies

Future commercial production may benefit from:

- Smart cultivation rooms
- Internet of Things (IoT) monitoring
- Robotic bottle handling
- Automated inoculation systems
- Cloud-based production management
- Blockchain-enabled traceability
- Digital twin technology for cultivation optimization

Scientific Terminology

Term	Meaning
Bioactive Compound	A naturally occurring substance with biological activity
Metabolomics	Study of metabolites in living organisms
Genomics	Study of the complete genetic material
Transcriptomics	Study of gene expression
Proteomics	Study of proteins
Nutraceutical	Food-derived product intended to support health

Practical Tips

✔ Follow peer-reviewed scientific literature rather than unsupported internet claims.

✔ Distinguish between laboratory evidence and clinical evidence.

✔ Use standardized cultivation methods to improve research reproducibility.

✔ Maintain complete experimental records.

✔ Collaborate with universities and accredited laboratories for advanced testing.

Case Study 22.1 – From Cultivation to Research Collaboration

A commercial cultivator partnered with a nearby university to analyze the cordycepin content of different production batches. By systematically recording substrate composition, incubation conditions, and harvest timing, the team identified cultivation variables associated with higher cordycepin concentrations. The collaboration resulted in improved production consistency and a jointly published research paper.

Lesson: Collaboration between industry and academia can accelerate innovation while improving product quality and scientific credibility.

Chapter Summary

This chapter explored the scientific foundation of *Cordyceps militaris*, including its principal bioactive compounds, analytical techniques, pharmacological research, biotechnology applications, and future research directions. While laboratory findings are promising, further clinical research is necessary to clarify many proposed health applications. Cultivators, researchers, and entrepreneurs who understand both the biological and scientific dimensions of this remarkable fungus will be better positioned to contribute to its responsible development.

Key Takeaways

✓ *Cordyceps militaris* contains numerous scientifically interesting bioactive compounds.

✓ Laboratory research is advancing rapidly, but clinical evidence remains limited for many proposed applications.

✓ Modern analytical methods support quality control and product standardization.

✓ Biotechnology, AI, and genomics are expected to transform future cultivation and research.

✓ Scientific integrity and evidence-based communication are essential for long-term success.

Chapter 23: Global Cultivation Practices, International Standards, Export Opportunities, and Commercial Case Studies

"The future of *Cordyceps militaris* is global. Scientific cultivation, international quality standards, and sustainable production are transforming this medicinal mushroom into one of the world's fastest-growing biotechnology crops."

Learning Objectives

After completing this chapter, you will be able to:

- Understand global cultivation trends.
- Compare cultivation technologies used in different countries.
- Learn international quality standards.
- Explore export opportunities.
- Study successful commercial production models.
- Identify future global market trends.
- Develop strategies for international

competitiveness.

23.1 Introduction

The cultivation of *Cordyceps militaris* has expanded rapidly over the last two decades. Once confined primarily to research laboratories, it is now cultivated commercially in several countries using advanced biotechnology and controlled-environment agriculture.

The growing global demand is driven by:

- Functional foods
- Nutraceuticals
- Herbal products
- Wellness supplements
- Scientific research
- Biotechnology applications

Countries investing in high-quality cultivation systems are increasingly supplying both domestic and international markets.

23.2 Global Market Overview

The international market for medicinal mushrooms has grown steadily due to increasing consumer interest in natural health products.

Major product categories include:

- Whole dried fruiting bodies
- Mushroom powders
- Capsules
- Extracts
- Functional beverages
- Nutraceutical ingredients

Demand is strongest in countries with established wellness and functional food markets.

23.3 Major Producing Countries

Commercial cultivation is reported in several countries.

China

China remains one of the largest producers of cultivated medicinal mushrooms.

Strengths include:

- Large-scale production
- Advanced biotechnology
- Extensive research
- Export-oriented industries

South Korea

Known for:

- High technology cultivation
- Automated production systems
- Premium-quality products
- Strong government research support

Japan

Focuses on:

- High-value products
- Quality control
- Precision cultivation
- Pharmaceutical research

Thailand

Strengths include:

- Tropical cultivation research
- Functional food development
- University-industry collaboration

Vietnam

Rapidly expanding commercial production.

Increasing focus on export-quality cultivation.

Taiwan

Well known for:

- Biotechnology
- Controlled-environment agriculture
- Commercial innovation

United States

Research focuses on:

- Functional foods
- Dietary supplements

- Biotechnology
- Mushroom extraction technologies

European Countries

Research emphasizes:

- Food safety
- Organic production
- Sustainable cultivation
- Standardized quality control

23.4 Comparison of Cultivation Technologies

Country	Major Strength
China	Large-scale commercial production
South Korea	Automation
Japan	Precision quality management
Taiwan	Biotechnology
Thailand	Functional food innovation
USA	Nutraceutical development
Europe	Food safety and sustainability

23.5 International Cultivation Trends

Current trends include:

- Automation
- AI monitoring
- IoT sensors
- Climate-controlled cultivation
- Robotic bottle handling
- Digital production management
- Precision agriculture

These technologies improve consistency and reduce labour requirements.

23.6 Sustainable Production

Modern cultivation emphasizes environmental responsibility.

Important practices include:

- Water conservation
- Energy-efficient lighting
- Renewable energy
- Waste recycling
- Reusable bottles
- Sustainable packaging

Sustainability increasingly influences purchasing decisions.

23.7 Circular Economy

Commercial laboratories are exploring methods to reuse or recycle production residues.

Examples include:

- Compost production
- Organic fertilizers
- Vermicomposting
- Bioenergy research

Spent substrate should be managed responsibly according to local environmental regulations.

23.8 International Quality Standards

Export markets often expect compliance with recognized quality systems.

Examples include:

- Good Manufacturing Practices (GMP)
- HACCP-based food safety systems

- ISO quality management standards
- Traceability systems

Specific requirements vary among countries and product categories.

23.9 Product Standardization

International buyers often request:

- Batch consistency
- Moisture analysis
- Microbiological testing
- Heavy metal analysis
- Pesticide residue testing (where applicable)
- Certificate of Analysis (COA)

Standardization increases buyer confidence.

23.10 Export Documentation

Export requirements vary by destination country.

Typical documentation may include:

- Commercial Invoice

- Packing List
- Certificate of Origin
- Phytosanitary or health-related documents (where applicable)
- Quality certificates
- Transport documents

Exporters should verify current import requirements with the destination country before shipment.

23.11 Cold Chain Considerations

Fresh products require temperature-controlled transportation.

Dried products generally require:

- Moisture protection
- Clean packaging
- Physical protection during transport

Proper logistics preserve product quality.

23.12 International Packaging

Export packaging should provide:

✔ Moisture resistance

✔ Tamper evidence

✔ Attractive appearance

✔ Barcode compatibility

✔ Batch traceability

✔ Shipping durability

23.13 E-commerce Opportunities

Global customers increasingly purchase medicinal mushrooms through:

- Company websites
- Online marketplaces
- Specialty wellness platforms
- International distributors

Professional product photography and accurate information improve customer confidence.

23.14 Research Collaboration

Collaboration between:

- Universities

- Biotechnology companies
- Government institutes
- Commercial farms

accelerates innovation and technology transfer.

Joint research often leads to:

- Improved strains
- Better cultivation methods
- Scientific publications
- Product innovation

23.15 Intellectual Property

Commercial innovation may involve:

- Trademarks
- Copyrights
- Patents
- Trade secrets
- Plant and microbial intellectual property considerations

Entrepreneurs should seek professional legal advice before filing intellectual property applications.

23.16 International Business Challenges

Challenges include:

- Competition
- Regulatory changes
- Quality expectations
- Shipping costs
- Currency fluctuations
- Market access
- Consumer education

Long-term success depends upon adaptability.

23.17 Global Opportunities

Emerging opportunities include:

- Functional nutrition
- Healthy ageing products
- Sports nutrition

- Biotechnology
- Precision fermentation
- Personalized wellness
- Premium gifting
- Medical research

23.18 Case Study – China

Large commercial enterprises use:

- Automated bottle filling
- Robotic inoculation
- Environmental monitoring
- Computerized production

Advantages:

- High productivity
- Uniform quality
- Reduced labour costs

23.19 Case Study – South Korea

Commercial farms emphasize:

- Controlled lighting

- Climate automation
- Digital monitoring
- Premium export products

Strong collaboration exists between universities and industry.

23.20 Case Study – India

India possesses excellent potential due to:

- Growing biotechnology sector
- Skilled workforce
- Increasing demand for wellness products
- Expanding e-commerce
- Government support for MSMEs and startups

Challenges include:

- Limited commercial-scale production
- Need for technical training
- Limited awareness among consumers

23.21 SWOT Analysis for Indian Entrepreneurs

Strengths	**Weaknesses**
Growing domestic market	Limited trained manpower
Lower labour costs	Limited advanced laboratories
Strong IT sector	Technology gap
Opportunities	**Threats**
Export potential	International competition
Functional food market	Regulatory changes
Startup ecosystem	Counterfeit products

23.22 International Business Strategy

A successful exporter should focus on:

✔ Consistent quality

✔ Scientific credibility

✔ Professional branding

✔ Customer education

✔ International certifications

✔ Digital marketing

✔ Strong distributor relationships

Commercial Growth Model

Research

|

▼

Pilot Production

|

▼

Commercial Production

|

▼

Quality Certification

|

▼

National Brand

|

▼

Export Business

|

▼

Global Recognition

Export Readiness Checklist

- ☐ Product standardized
- ☐ Laboratory tested
- ☐ Packaging approved
- ☐ Labels compliant
- ☐ Documentation prepared
- ☐ Product traceable
- ☐ Shipping partner selected
- ☐ Customer support established

Future Technologies

Emerging technologies likely to shape the next decade include:

- Artificial Intelligence
- Machine Vision
- Robotics
- Blockchain Traceability
- Internet of Things (IoT)
- Smart Greenhouses
- Automated Environmental Control
- Digital Quality Management Systems

Practical Tips

✔ Study successful international producers.

✔ Continuously improve cultivation techniques.

✔ Participate in scientific conferences.

✔ Collaborate with universities.

✔ Maintain international quality standards.

✔ Invest in research and development.

✔ Build long-term customer relationships.

Case Study 23.2 – A Small Laboratory Becomes an Export Brand

A small biotechnology laboratory began with only 500 cultivation bottles per production cycle. Rather than competing on volume, the company focused on premium quality, complete batch traceability, and scientifically documented cultivation practices. It invested in attractive packaging, multilingual labels, and a strong digital presence. Within a few years, the laboratory established partnerships with overseas distributors interested in specialty mushroom products. Its success demonstrated that consistent quality and professional management can create export opportunities even for relatively small producers.

Lesson: International competitiveness depends more on quality, credibility, and consistency than on production volume alone.

Chapter Summary

This chapter explored the global landscape of *Cordyceps militaris* cultivation, including production practices in different countries, international quality expectations, export considerations, sustainability, and emerging technologies. It highlighted the importance of standardization, innovation, and collaboration in building globally competitive mushroom enterprises.

Key Takeaways

✓ *Cordyceps militaris* is now a globally recognized biotechnology crop.

✓ International buyers expect consistent quality, traceability, and regulatory compliance.

✓ Automation, AI, and precision cultivation are shaping the future of production.

✓ India has significant potential to become an important producer with investment in technology and quality systems.

✓ Sustainable practices and continuous innovation are essential for long-term success.

Chapter 24: Frequently Asked Questions (FAQs)

Practical Troubleshooting Guide, Standard Operating Procedures (SOPs), Ready-to-Use Record Formats, and Commercial Checklists

"Every successful cultivator encounters problems. The difference between a beginner and an expert lies not in avoiding mistakes, but in identifying, correcting, and preventing them through scientific methods."

Learning Objectives

After completing this chapter, you will be able to:

- Solve common cultivation problems.
- Diagnose contamination and growth disorders.
- Use practical Standard Operating Procedures (SOPs).
- Maintain professional laboratory records.
- Improve production consistency.
- Reduce crop losses.
- Operate a commercial Cordyceps

laboratory efficiently.

PART I

Frequently Asked Questions (FAQ)/Section A – General Questions

Q1. What is *Cordyceps militaris*?

It is a cultivated medicinal mushroom widely researched for its bioactive compounds, especially cordycepin, and used in functional foods, nutraceuticals, and scientific research.

Q2. Is *Cordyceps militaris* the same as *Ophiocordyceps sinensis*?

No.

Although both belong to the Cordyceps group, they are different species with different biology, ecology, and cultivation methods.

Q3. Can beginners cultivate Cordyceps?

Yes.

With proper laboratory hygiene, training, and environmental control, beginners can successfully cultivate it on a small scale.

Q4. Is a laminar airflow cabinet necessary?

For commercial production, it is strongly recommended because aseptic inoculation greatly reduces contamination risk.

Q5. Can I cultivate Cordyceps at home?

Yes.

A dedicated clean room or small laboratory with environmental control is preferable to maintain consistent results.

Q6. Which strain should I purchase?

Obtain authenticated cultures from reputable culture collections, universities, or recognized laboratories. Avoid unknown or repeatedly subcultured strains of uncertain origin.

Q7. How long can a culture be maintained?

Master cultures should be preserved appropriately (e.g., refrigerated or cryopreserved by specialized laboratories). Working cultures should be renewed periodically from verified stock cultures to maintain vigor.

Q8. Can old cultures produce good yields?

Repeated subculturing may reduce vigor in some strains. Periodic replacement from a healthy master culture is recommended.

Q9. Which substrate gives the highest yield?

There is no universal substrate. Yield depends on strain, formulation, sterilization, environmental conditions, and cultivation management. Optimization should be based on trials.

Q10. Why is rice commonly used?

Rice provides a convenient and consistent cereal substrate for laboratory cultivation and has been widely used in research and commercial production.

Section B – Media Preparation

Q11. Why is my substrate too wet?

Possible causes:

- Excess water
- Inadequate mixing
- Incorrect measurement

Solution:

✔ Measure water accurately.

✔ Mix thoroughly before bottling.

Q12. Why is the substrate too dry?

Possible causes:

- Insufficient water
- High evaporation during preparation

Solution:

Adjust water addition carefully and standardize the preparation procedure.

Q13. Can tap water be used?

Potable water may be suitable in some situations, but purified or filtered water provides greater consistency and reduces potential contaminants.

Q14. Can I reuse old substrate?

No.

Fresh substrate should be prepared for each production cycle.

Q15. Can I sterilize twice?

Repeated sterilization may alter substrate characteristics. Validate sterilization procedures to minimize the need for reprocessing.

Section C – Sterilization

Q16. Why do bottles become contaminated after sterilization?

Possible reasons:

- Incomplete sterilization
- Leaking bottle closures
- Contaminated inoculum
- Poor aseptic technique

Q17. How often should the autoclave be serviced?

According to the manufacturer's recommendations and laboratory maintenance schedule. Routine inspection and calibration are essential.

Q18. How can sterilization efficiency be verified?

By following validated sterilization procedures and, where appropriate, using biological or chemical indicators according to laboratory protocols.

Section D – Inoculation

Q19. Why does contamination occur immediately after inoculation?

Common causes:

- Dirty laminar airflow cabinet
- Poor hand hygiene
- Contaminated tools
- Airborne contamination
- Contaminated culture

Q20. Should I wear gloves?

Yes.

Clean disposable gloves reduce contamination risk when used correctly.

Q21. Should I disinfect forceps between bottles?

Yes.

Sterilize or disinfect tools according to laboratory SOPs before each transfer.

Q22. Can two strains be inoculated in one bottle?

This is generally not recommended unless conducting controlled research on mixed cultures.

Q23. Why is my mycelium not growing?

Possible reasons include:

- Non-viable culture
- Incorrect incubation temperature
- Excess substrate moisture
- Contamination
- Nutrient imbalance

Q24. Why is growth uneven?

Possible causes:

- Uneven inoculation
- Variable substrate composition
- Environmental differences

Q25. Should contaminated bottles be opened?

No.

Seal, isolate, and dispose of contaminated bottles according to biosafety procedures.

PART II

Troubleshooting Guide

Problem	Possible Cause	Solution
No mycelial growth	Poor culture viability	Replace with verified culture
Slow colonization	Environmental conditions not optimized	Review incubation parameters
Green mould	Contamination	Isolate and discard affected bottles
Black mould	Contamination	Improve aseptic practices
Wet substrate	Excess water	Standardize moisture content
Dry substrate	Insufficient moisture	Review substrate preparation
Uneven fruiting	Environmental variability	Improve climate control
Thin fruiting bodies	Environmental or nutritional factors	Optimize cultivation conditions
Poor colour	Inadequate	Review lighting

Problem	Possible Cause	Solution
	lighting or strain variation	program
Bottle cracking	Thermal shock	Allow gradual cooling after sterilization
Condensation	Temperature fluctuations	Stabilize incubation environment
Low yield	Multiple factors	Review entire production process

PART III

Standard Operating Procedures (SOPs)

SOP-1

Entering the Laboratory

1. Wear a clean laboratory coat.
2. Remove outdoor footwear if required.
3. Wash and sanitize hands.
4. Wear gloves, hair cover, and face mask.

5. Record entry in the laboratory logbook.

SOP–2

Laminar Air Flow Cabinet

Before use:

✔ Switch on the blower.

✔ Clean work surfaces with an appropriate disinfectant.

✔ Arrange sterilized materials before beginning work.

After use:

✔ Remove waste.

✔ Clean surfaces.

✔ Record usage if required.

SOP–3

Bottle Preparation

1. Inspect bottles.
2. Wash thoroughly.
3. Dry completely.
4. Fill substrate uniformly.
5. Close with sterile closures.

6. Label each bottle.

SOP–4

Harvesting

1. Inspect bottles.
2. Remove contaminated bottles.
3. Harvest mature fruiting bodies.
4. Transfer to clean trays.
5. Move promptly for drying.

PART IV

Ready-to-Use Record Formats/Daily Laboratory Register

Date	Operator	Room Temperature	Humidity	Remarks

Culture Register

Culture Code	Source	Date Received	Storage Location	Remarks

Sterilization Register

Date	Autoclave No.	Batch No.	Start Time	End Time	Operator

Inoculation Register

Date	Batch No.	Number of Bottles	Culture Used	Operator

Harvest Register

Date	Batch No.	Bottles Harvested	Yield (g)	Remarks

Drying Register

Batch No.	Drying Method	Start Date	Completion Date	Remarks

Packaging Register

Date	Product	Batch No.	Quantity Packed	Operator

Customer Complaint Register

Date	Complaint	Batch No.	Investigation	Action Taken

PART V

Daily Production Checklist/ Before Starting Work

☐ Laboratory cleaned

- ☐ Equipment checked
- ☐ Environmental conditions recorded
- ☐ PPE available
- ☐ SOPs accessible
- ☐ Raw materials ready

During Production

☐ Bottles inspected

- ☐ Labels verified
- ☐ Sterility maintained
- ☐ Contamination monitored
- ☐ Records updated

End of Day

☐ Waste disposed

- ☐ Equipment cleaned
- ☐ Laboratory disinfected
- ☐ Environmental records completed
- ☐ Doors secured

PART VI

Emergency Response Guide

Power Failure

✔ Keep culture rooms closed to preserve temperature.

✔ Activate backup power if available.

✔ Record the duration of the outage.

Water Supply Failure

✔ Suspend substrate preparation if adequate water is unavailable.

✔ Use stored purified water for essential laboratory work.

Major Contamination Event

✔ Isolate affected batches.

✔ Suspend inoculation if necessary.

✔ Sanitize the affected area.

✔ Investigate the root cause.

✔ Implement corrective and preventive actions (CAPA).

Case Study 24.1 – Learning from Contamination

A new cultivation laboratory experienced contamination in nearly 30% of its bottles during the first month. After reviewing records, the team discovered that forceps were not being adequately sterilized between inoculations. The laboratory revised its SOP, retrained personnel, and introduced a contamination log. Within the next three production cycles, contamination declined to below 5%, and production became more consistent.

Lesson: Careful documentation, staff training, and adherence to SOPs can dramatically improve production outcomes.

Chapter Summary

This chapter serves as a practical field manual for growers, researchers, and commercial producers. It combines frequently asked questions, troubleshooting guidance, SOPs, record formats, and operational checklists into a single reference designed for day-to-day laboratory use. Proper implementation of these tools helps reduce contamination, improve consistency, strengthen documentation, and support professional cultivation practices.

Key Takeaways

✓ Most cultivation problems can be prevented through disciplined laboratory practices.

✓ Written SOPs promote consistency and reduce human error.

✓ Comprehensive records improve traceability and decision-making.

✓ Regular review of production data supports continuous improvement.

✓ Troubleshooting should focus on identifying root causes rather than symptoms.

Chapter 25: The Future of Cordyceps militaris

Entrepreneurship, Innovation, Resources, Glossary, and Appendices

"Every successful cultivation laboratory begins with curiosity, grows through discipline, and flourishes through innovation. The future belongs to cultivators who combine scientific knowledge with ethical business practices and a commitment to continuous learning."

Learning Objectives

After completing this chapter, you will be able to:

- Understand future opportunities in the Cordyceps industry.
- Build a long-term biotechnology enterprise.
- Apply principles of innovation and sustainability.
- Identify research and collaboration

opportunities.

- Access recommended resources for further learning.
- Use the appendices and reference materials provided in this book.

25.1 Introduction

The journey of *Cordyceps militaris* from a little-known medicinal fungus to a globally cultivated biotechnology crop demonstrates the remarkable progress achieved through scientific research and technological innovation.

Today, *Cordyceps militaris* is cultivated not only by researchers but also by commercial entrepreneurs, biotechnology companies, universities, nutraceutical manufacturers, and controlled-environment agriculture specialists.

As global demand for functional foods and scientifically validated natural products continues to increase, *Cordyceps militaris* is expected to remain one of the most promising medicinal mushrooms of the twenty-first century.

25.2 Vision for the Future

The future of Cordyceps cultivation will be characterized by:

- Precision agriculture
- Artificial Intelligence
- Robotics
- Genomics
- Climate-smart cultivation
- Sustainable biotechnology
- Personalized nutrition
- Digital agriculture

Future laboratories may become almost fully automated, with real-time environmental monitoring, predictive analytics, and AI-assisted quality control.

25.3 Entrepreneurship Opportunities

The Cordyceps industry offers diverse entrepreneurial possibilities.

Potential business areas include:

Production

- Fresh mushrooms
- Dried mushrooms
- Spawn production
- Culture laboratory

Processing

- Powder
- Capsules
- Tea
- Extracts
- Functional foods

Biotechnology

- Tissue culture
- Strain improvement
- Contract research
- Laboratory services

Education

- Training institutes

- Online courses
- Demonstration farms
- Consultancy

Digital Business

- E-commerce
- Export
- Technical publications
- Educational videos

25.4 Skills Required for Future Entrepreneurs

A successful entrepreneur should continuously develop skills in:

✔ Mushroom biology
✔ Laboratory techniques
✔ Quality management
✔ Business planning
✔ Marketing
✔ Digital communication
✔ Financial management
✔ Leadership
✔ Regulatory compliance

✔ Customer relationship management

25.5 Innovation Roadmap

Innovation may involve:

- Improved strains
- Sustainable substrates
- Better environmental control
- Eco-friendly packaging
- Smart sensors
- Automated harvesting
- Digital traceability
- Advanced extraction technologies

Continuous innovation is essential for remaining competitive in the global market.

25.6 Sustainability

Sustainable cultivation should focus on:

- Reduced energy consumption
- Renewable energy
- Efficient water use
- Recyclable packaging

- Responsible waste management
- Carbon footprint reduction

Environmental responsibility is becoming an important factor in consumer purchasing decisions.

25.7 Ethical Business Practices

A successful enterprise should always emphasize:

- Honest product information
- Scientific accuracy
- Transparent labeling
- Fair pricing
- Customer education
- Respect for intellectual property
- Environmental stewardship

Avoid exaggerated health claims that are not supported by reliable scientific evidence or permitted under applicable regulations.

25.8 Building a Scientific Brand

A respected brand is built through:

- Consistent product quality
- Transparent documentation
- Professional packaging
- Customer trust
- Continuous improvement
- Scientific publications
- Educational outreach

Trust takes years to build and only moments to lose.

25.9 Collaboration

Long-term success is often achieved through collaboration with:

- Universities
- Biotechnology institutes
- Government agencies
- Farmer Producer Organizations (FPOs)
- Start-up incubators
- International research networks

Partnerships facilitate knowledge exchange and technological advancement.

25.10 Lifelong Learning

Cultivators should remain updated through:

- Scientific journals
- Professional workshops
- Conferences
- Webinars
- Research collaborations
- Continuing education programmes

Science evolves rapidly; successful entrepreneurs evolve with it.

25.11 Future Research Priorities

Important research areas include:

- High-yielding strains
- Bioactive compound enhancement
- Climate-resilient cultivation
- Precision fermentation
- AI-assisted production

- Automated contamination detection
- Sustainable substrates
- Shelf-life optimization
- Functional food innovation

25.12 Advice for Beginners

Start small.

Master laboratory discipline.

Maintain detailed records.

Do not rush into large-scale production before gaining practical experience.

Reinvest profits into quality improvement rather than rapid expansion.

25.13 Advice for Commercial Producers

Focus on:

✔ Quality

✔ Consistency

✔ Documentation

✔ Innovation

✔ Customer satisfaction

✔ Regulatory compliance

✔ Staff training

✔ Brand development

25.14 Advice for Researchers

Maintain scientific integrity.

Publish reproducible research.

Collaborate across disciplines.

Translate laboratory discoveries into practical cultivation improvements.

25.15 Advice for Students

Develop both theoretical and practical skills.

Seek internships.

Visit commercial laboratories.

Participate in scientific conferences.

Remain curious.

25.16 A Message to Policymakers

Support for medicinal mushroom cultivation can contribute to:

- Biotechnology development
- Rural entrepreneurship
- Employment generation
- Value-added agriculture
- Export growth

- Scientific innovation

Public-private partnerships can accelerate industry development.

25.17 A Message to Readers

Whether you are:

- A student,
- A researcher,
- A farmer,
- A biotechnology entrepreneur,
- A commercial producer,

remember that excellence is achieved through discipline, observation, and continuous improvement.

Every successful cultivation cycle begins with careful preparation and ends with thoughtful evaluation.

Final Commercial Checklist

Before launching a commercial venture:

☐ Technical training completed

☐ Laboratory established

☐ SOPs prepared

☐ Quality system implemented

☐ Trial batches successful

☐ Branding finalized

☐ Packaging designed

☐ Marketing strategy developed

☐ Financial planning completed

☐ Regulatory requirements verified

Recommended Reading

Readers are encouraged to consult:

- Peer-reviewed journals on mycology and biotechnology
- Books on medicinal mushrooms
- Food safety and quality management references
- Biotechnology laboratory manuals
- Scientific publications on *Cordyceps militaris*

Always prioritize recent, peer-reviewed sources over unverified online information.

Glossary (Selected Terms)

Term	Definition
Agar	A gelatinous medium used for microbial culture.
Aseptic Technique	Procedures used to prevent contamination.
Autoclave	Equipment used for sterilization with steam under pressure.
Batch	A defined quantity produced under similar conditions.
Bioactive Compound	A naturally occurring substance with biological activity.
Cordycepin	A characteristic nucleoside analogue found in *Cordyceps militaris.*
Fruiting Body	The visible reproductive structure of the fungus.
HEPA Filter	High-Efficiency Particulate Air filter used to remove airborne particles.
Mycelium	The vegetative network of fungal hyphae.
Spawn	Fungal inoculum used for cultivation.

Term	Definition
SOP	Standard Operating Procedure.
Traceability	Ability to track a product through all stages of production.

Useful Conversion Table

Measurement	Equivalent
1 kilogram	1000 grams
1 litre	1000 millilitres
1 metre	100 centimetres
1°C	33.8°F
1 inch	2.54 cm

Appendix A – Essential Laboratory Equipment Checklist

- Laminar Air Flow Cabinet
- Autoclave
- pH Meter
- Analytical Balance
- Refrigerator
- Incubator
- Hygrometer
- Thermometer
- LED Lighting
- Vacuum Packaging Machine
- Drying Unit
- Laboratory Glassware
- Stainless Steel Tools
- Personal Protective Equipment (PPE)

Appendix B – Production Records

The following records should be maintained:

- Culture Register
- Spawn Register
- Batch Manufacturing Record
- Environmental Monitoring Log
- Equipment Maintenance Register
- Harvest Register
- Drying Register
- Packaging Register
- Sales Register
- Customer Complaint Register

Appendix C – Laboratory Safety Rules

1. Wear appropriate PPE.
2. Maintain laboratory cleanliness.
3. Sterilize equipment before and after use.
4. Handle cultures aseptically.
5. Dispose of biological waste safely.
6. Record accidents immediately.
7. Keep emergency contact numbers available.
8. Restrict unauthorized access.

Appendix D – Suggested Practical Exercises

1. Preparation of agar media.
2. Pure culture isolation.
3. Mycelial growth comparison on different media.
4. Substrate preparation.
5. Bottle sterilization.
6. Aseptic inoculation.
7. Environmental monitoring.
8. Harvesting and drying.
9. Packaging and labeling.
10. Cost analysis of a production batch.

Final Case Study – From Curiosity to Commercial Success

A retired professional with a lifelong interest in science began cultivating *Cordyceps militaris* in a modest laboratory. Rather than aiming for immediate commercial success, the cultivator invested time in understanding fungal biology, documenting each production cycle, and continuously refining laboratory practices. Over several years, the laboratory expanded into a respected enterprise producing high-quality cultivated mushrooms, educational materials, and training programs for aspiring growers.

The key factors behind this success were:

- Careful observation and documentation.
- Consistent quality standards.
- Continuous learning.
- Transparent communication with customers.
- Gradual expansion based on experience rather than speculation.

Lesson: Sustainable success in biotechnology is built on patience, scientific thinking, and an unwavering commitment to quality.

Final Summary of the Book

This has presented a comprehensive guide to *Cordyceps militaris*, covering its biology, taxonomy, laboratory establishment, culture maintenance, substrate preparation, sterilization, inoculation, incubation, fruiting, harvesting, post-harvest handling, quality assurance, value addition, economics, commercialization, research, and global trends. It has also provided practical tools, troubleshooting guides, SOPs, record formats, and business planning strategies to support readers at every stage of their cultivation journey. The aim of this book is not merely to teach cultivation techniques but to encourage readers to adopt a scientific, ethical, and professional approach to medicinal mushroom production.

Final Key Takeaways

✓ Build your enterprise on science, not myths.

✓ Prioritize quality over quantity.

✓ Maintain complete documentation and traceability.

✓ Continue learning from research and practical experience.

✓ Innovate responsibly while complying with applicable regulations.

✓ Share knowledge and collaborate with the scientific community.

✓ Remember that sustainable success comes from consistency, integrity, and a commitment to continuous improvement.

CORDYCEPS MILITARIS CULTIVATION: A PRACTICAL GUIDE.

PART-III

Diagram & Machines Image Useful for Cultivation of Cordyceps militaries

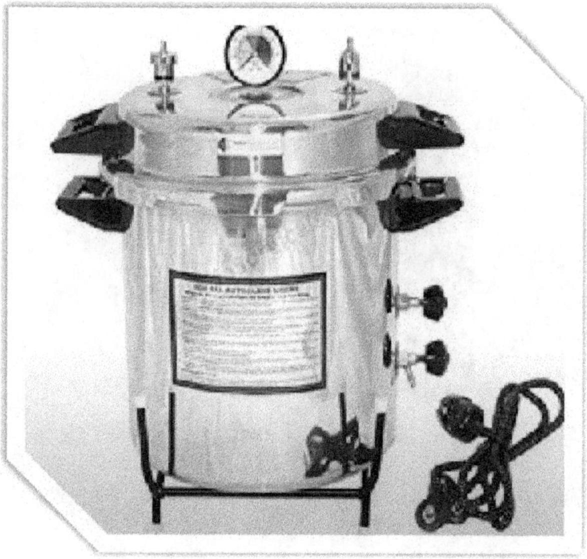

Autoclave

Laminor

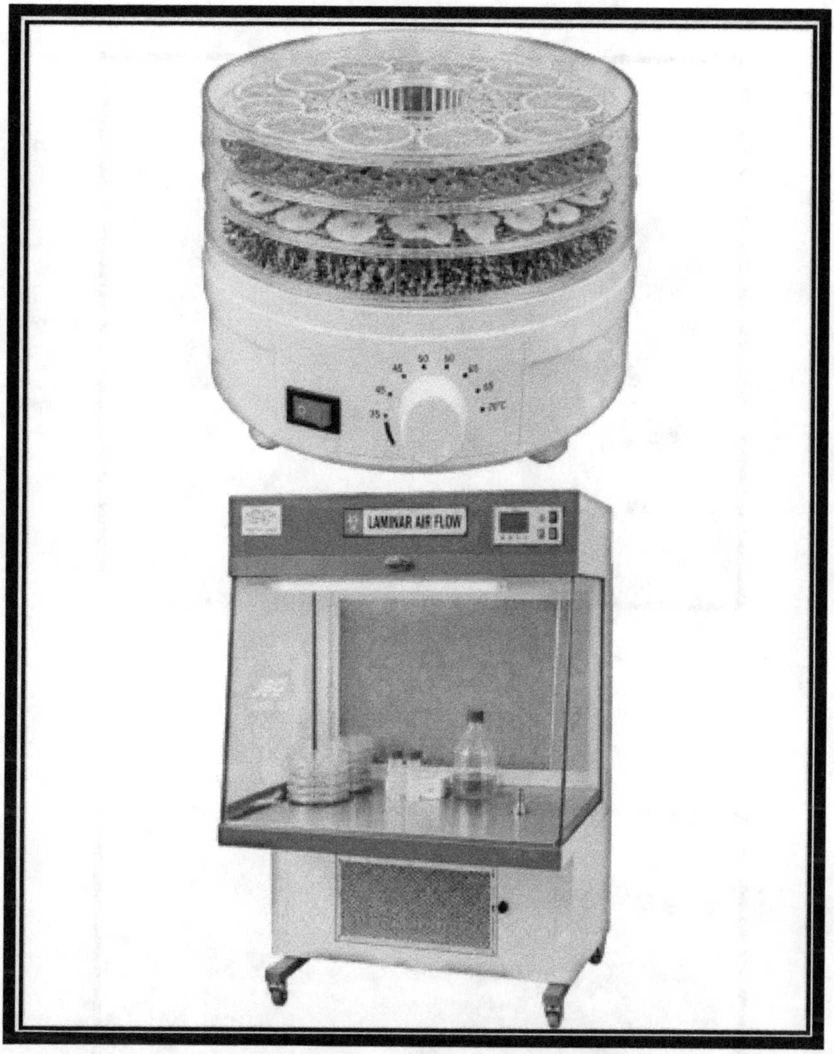

Dehyder & Laminor

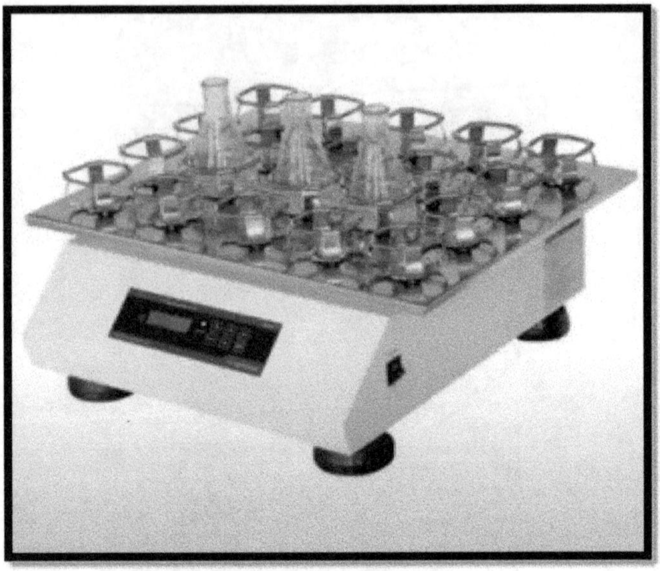

Rotary Shaker

Rack & Bekar

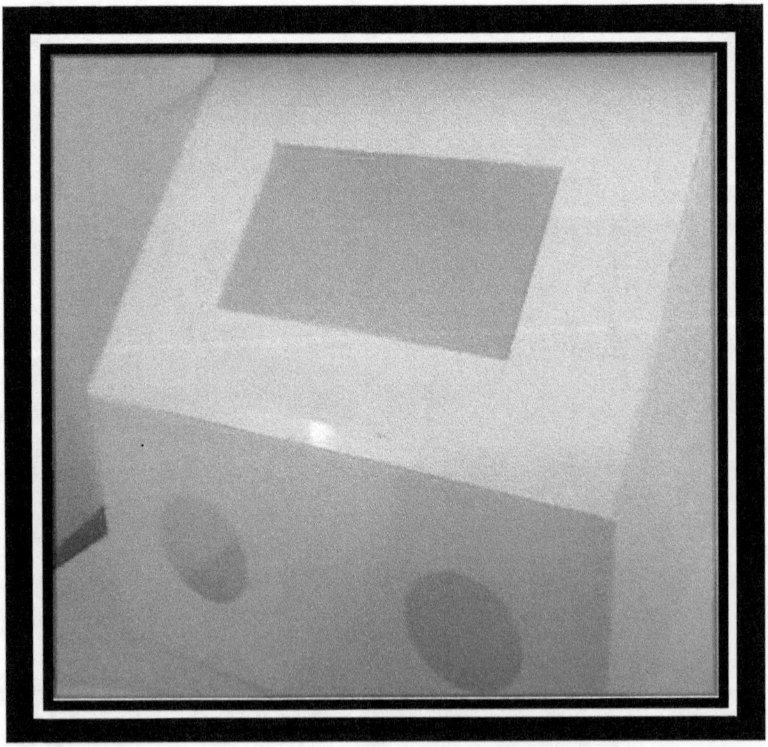

कोडिसेप्स की घेरलू खेती के लिए संपूर्ण सिस्तम

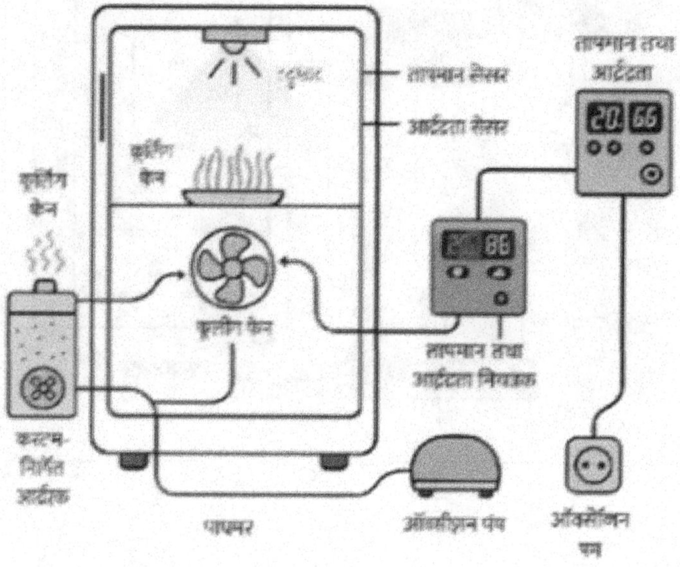

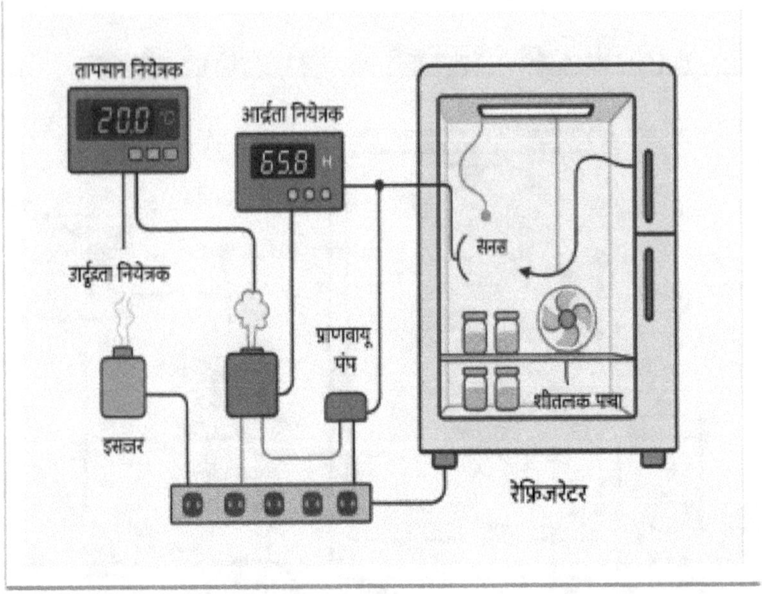
तापमान नियेत्रक
20.0 °C
आर्द्रता नियेत्रक
65.8 H
प्राणवायू
पंप
रेफ्रिजरेटर

Did you love *Cordyceps militaris Cultivation: A Practical Guide.*? Then you should read *Medical Astrology: Insights into Health, Destiny, and the Future*[1] by Pradip Kumar Ray!

[2]

Read more at https://pkrburpub.com.

1. https://books2read.com/u/4DVQzr
2. https://books2read.com/u/4DVQzr

Also by Pragyan Ray

Cordyceps militaris Cultivation: A Practical Guide.

Also by Pradip Kumar Ray

Eternal Witnesses
Kumbh Mela Chronicles
Siri Bhoovalay : A Miraculous Book
A Journey through Heart and Mind
Dashavatara (The Ten Incarnations of Lord Vishnu and Their Purpose)
Life, Strategy, Human Spirit (The Relevance of Mahabharata Today)
Divine Echoes: The Living God of Today (The Sacred Prabachans of Baba Premananda Ji Maharaj)
Galactic Horizons (Tales of Discovery, Power & Space)
Two Tales of Triumph (From Village Fields to City Dreams)
Boundless Hearts (From Medicine to Marriage, Stories of Strength)
Chicken Neck of WB (India's Strategic Lifeline)

The Truth Behind POK (History, Conflict, and the Path Forward)

Living Joyfully Beyond Sixty (Timeless Ways to Happiness from 60 to 80 and Beyond)

India's Steel Resolve (Army, Navy, and Air Force in Action)

The Unseen Threads(Karma, Compassion, and the Killing of Animals)

Smiles Behind the Curtain of Power (The Silent Burden of a Prime Minister)

Strategies of Power (Amit Shah's Leadership in Indian Politics)

The Architect of New India (Modi's Vision, Power, and Legacy)

Mamata Banerjee (A Beacon of Women Leadership in Indian Politics)

Shyama Prasad Mukherjee (The Forgotten Architect of Modern India)

Timeless Teaching of the Bhagvad Gita on Human Psychology

Bengali Dishes: Classic Modern Twists

Chinmoy Maharaj (Voices of Unity for Hindu Rights in Bangladesh)

Shadows of Power (The Conspiracy behind Hasina's Downfall)

Ajit Doval (The Architecture of Modern India's Security)

Durga Puja Narrative : A Celebration in Bengal

Medical Astrology: Insights into Health, Destiny, and the Future

Dad Means Dad (The Unexplored Hero)

Mom means Mom (The Unconditional Eternal Love)

Melody of a Nation: Life and Legacy of Lata Mangeshkar

Ratan Tata: Architect of Modern India

Mithun Chakraborty: A Hero's Path to the Dada Saheb Phalke Award

The Inherent Meaning of Mythology

Legends of Ramayana: A Timeless Epic

Unknown Facts of Epic Ramayana

Exploring Vedic Teachings (Insights for Everyday Life and Spiritual Growth)

Exploring Vedic Thought (From Knowledge to Liberation)

Subhendu Adhikari (Biography of a Struggling Politician)
The National Presidents of the Bharatiya Janata Party [Leadership, Organisation, Ideology and Historical Evolution (1980–Present)]
Transform Your Thinking, Transform Your Life (Inspiring Stories to Change Your Mind-Set)
Cordyceps militaris Cultivation: A Practical Guide.

www.ingramcontent.com/pod-product-compliance
Lightning Source LLC
LaVergne TN
LVHW010710061126
851857LV00014B/1123
* 9 7 9 8 2 3 5 4 1 9 8 9 6 *